OUT OF THIS WORLD

"Look!" Mike shouted. "Here it comes!"

Stretching down from the highest clouds, the beautiful, shimmering Rainbow reached closer and closer. Finally it touched the ground right in front of Mike and his friends.

"Amazing," Steve murmured.

Mike's dog, Mutt, darted forward and jumped into the bands of colour. Multicoloured sparks leapt from the Rainbow and flew around the whole group.

Now Mike and his friends were inside the Rainbow.

Gently it lifted them off the ground. Mike looked up and saw a whole ocean of dazzling colour flowing before him. He and his friends were being pulled higher and higher into the sky!

RAINBOW

By Caroline Beecham

Based on the screenplay by
Ashley Sidaway and Robert Sidaway

RED FOX

This book is dedicated to Paul Anthony Gill

A Red Fox Book

Published by Random House Children's Books
20 Vauxhall Bridge Road, London SW1V 2SA

A division of Random House UK Ltd
London Melbourne Sydney Auckland
Johannesburg and agencies throughout the world

First published in the USA by Random House, Inc., New York
and simultaneously in Canada by Random House of Canada Limited, Toronto 1996

Red Fox edition 1996

Photographs by Yves Charbonneau

1 3 5 7 9 10 8 6 4 2

Phototypeset in Ehrhardt by Intype London Ltd

Printed and bound in Great Britain by
Cox & Wyman Ltd, Reading, Berkshire

Papers used by Random House UK Ltd are natural, recyclable products made from wood grown in sustainable forests. The manufacturing processes conform to the environmental regulations of the country of origin.

RANDOM HOUSE UK Limited Reg. No. 954009

ISBN 0 09 971301 2

1

Mike Bailey pedalled fast across the bare lot and into the railroad yard. He was already late and now the clouds above him looked heavy with rain.

The skyscrapers of Hudson Harbour, New Jersey were just ahead. Maybe he could beat the downpour if he ignored the burning sensation in his lungs and kept his mountain bike moving.

He weaved between the abandoned graffiti-covered railcars, dodging the web of train tracks that led towards the city. He didn't notice the furry face watching him from inside one of the old freight cars. He just kept pedalling hard until he came to a stop outside an old apartment house.

As Mike chained his bike to a drainpipe, he heard muffled applause. He smiled as he rushed downstairs to his grandfather's apartment. He had made it in time after all.

A flimsy paper sign on the apartment door read FRANK BAILEY: THE MAGIC MAN. Mike let himself in quietly. The furniture in the small living room had been pushed to one side. There was enough space for twenty cramped seats, but only half were taken. All the spectators were older people.

At the front of the room, Mike's grandfather, Frank,

was well under way with his act. He looked like a wizard from a storybook. A flowing purple satin cloak was draped around him, and his long grey hair stuck out from beneath a red cone-shaped hat. He was almost as old as the men and women in his audience, but his eyes twinkled like a young man's. Frank was a man who loved being alive.

As Mike unbuckled his bike helmet, he studied the large canvas poster hanging on the wall. Huge letters across the top proclaimed: THE AMAZING BAILEY AND SON. The picture showed a motorcyclist – Mike's father – jumping through a Wall of Fire. Mike had seen the poster a hundred times before, but it always made him sad.

Mike was eleven years old, and he had barely known his father. The man had taken off when Mike was only three. No note, no reasons. Nothing.

"Ladies and gentlemen, observe!" cried Frank as he lit a small torch and waved it over a top hat. Mike heard someone in the audience snoring loudly.

"I will burn a hole in the astral plane, allowing the rabbit to leap through a portal in time!" Frank announced, putting a fake rabbit inside the top hat. But when he pulled it back out again it was a *real* rabbit!

The audience applauded – or at least the ones who were still awake did.

"Now ladies and gentlemen . . ." Frank continued.

Mike rubbed his eyes. For some reason they were beginning to sting. Then Mike realized the air was growing smoky. Had his grandfather forgotten something on the stove? Mike began to scan the apartment looking for the source of the smoke.

Suddenly an old man leapt up and yelled, "Fire!"

Mike looked where the man was pointing – towards the front of the room. Frank's *cloak* was smoking. Flames were beginning to lick their way up his clothing. His grandfather was the thing on fire!

An instant later the room was in chaos. Chairs toppled over as people rushed for the door. Screams and shouts bounced off the walls of the small apartment.

Mike rushed into the kitchen and grabbed the fire extinguisher. He pulled out the pin as he tore back to the living room. The chairs were empty and the front door stood half open. Frank was frantically beating at the flames when Mike aimed the fire extinguisher nozzle and pressed the lever.

Whoosh! A second later the fire was out and Frank was covered with white foam.

"Are you okay?" Mike asked his grandfather.

Frank nodded and then Mike started giggling – he couldn't help it, his grandfather looked so funny – like an ice-cream sundae with extra whipped cream.

His grandfather started chuckling next. Soon both of them were doubled over in a fit of laughter.

By the time they stopped, a few minutes later, their sides were aching.

After Mike helped clean up the mess, he told his grandfather, "Well, I better get going."

"Sure you don't want to hang around awhile?" his

grandfather asked as Mike picked up his bike helmet from the small table near the door.

"Mom's expecting me," Mike said, looking down at his shoes.

Frank nodded sadly. He understood what Mike didn't have the heart to say – Mike's mother didn't like Mike to spend time with his grandfather. They both knew why. His mom blamed Frank for her husband leaving.

Frank looked out of the window in silence. It was drizzling.

"I'll give you a ride," he offered. "My van's right outside."

"No, I can ride my bike," Mike said. "Besides, Mom . . . you know . . ." Mike let his voice trail off.

"Okay," Frank said. "Well, thanks for helping with my grand finale. Maybe I should keep it as part of my act."

Mike was smiling as he left his grandfather's apartment. There was never a dull moment around Frank Bailey, that was for sure.

It was raining hard by the time Mike reached the railroad yard. But Mike had no trouble seeing where he was going. The sun was still shining brightly. Weird, Mike thought as he sped along beside the unused tracks.

Then something caught Mike's eye. A dog, a mangy-looking one. It was running alongside him. Suddenly it turned and ran straight at him. The bike's brakes squealed in protest as Mike tried to avoid hitting the animal.

He nearly tumbled over the handlebars as the bike screeched to a stop, inches from the mutt.

That was close, Mike thought as he got off his bike. The dog was barking its head off and wagging its tail happily.

"What is it, boy?" Mike asked.

The dog kept barking and darting back and forth. Mike examined it more closely. It didn't look like any dog he had ever seen before. And there was something strange about it that went beyond its looks. Mike could have sworn that the animal was trying to tell him something.

"You want me to follow you? Is that it?" Mike asked.

Suddenly, as if in answer to Mike's question, the dog bolted away.

"Okay, I'm coming." Mike ran after the dog and saw it disappear under an old boxcar covered with multi-coloured graffiti.

He heard a slight tinkling noise, like wind chimes and felt a gentle wind breathe over him. Gradually the sound grew stronger and changed to a hum.

Suddenly the purest, most brilliant pulsating light shot down from the sky in a fantastic Rainbow. Colours danced around Mike, reflecting off his body and the surface of the steel car behind him. Like fireworks, the sparkling currents bounced off the metal and shattered glass in a symphony of colours.

Mike barely noticed the dog standing at his side. He was mesmerised by the fantastic light show.

The low rumbling hum grew louder. The colours

ebbed and shimmered and a sudden rush of air rocked Mike on his heels. Then, in a flash, the colours swirled up his body and the Rainbow shot away as suddenly as it had appeared!

2

Mike crashed through his apartment door and into the living room. He couldn't wait to tell his mom and brother about the incredible Rainbow he'd seen.

Mike found his older brother in front of the TV – as usual. "Steve, you won't believe what I – "

Steve screwed up his face as he looked Mike over. "What'd you do, swim here?'

"No, I – "

"Get outta here," his brother shoved him away with his feet. "You're dripping on me."

Maybe Mom will be interested, Mike thought as he rushed into the kitchen. Jackie glanced up from the stove when she saw him run in. "Hey, sweet pea," she said as she pushed her long auburn hair back from her face. "I didn't hear you come in. How come you're so late? Dinner's almost ready."

Mike could see that his mother had just got home herself. She was still dressed in her work clothes – she was an assistant producer at the local TV news station.

"I saw something. Something really fantastic – "

"That's nice," said his mom, but she was hardly listening. She'd already turned from Mike to carry their supper out of the kitchen. "Steven," she called into the living room, "please sit at the table like a civilised human being!"

Why won't they pay attention to me, Mike thought as he went to his bedroom to change out of his wet clothes.

The next morning sunlight streamed through his bedroom window. Mike opened his eyes and the first thing he saw was the dinosaur poster on his wall.

Something was different about it, he thought. The primary colours in it seemed to be leaping out at him. The dinosaurs looked almost as if they were alive!

Strange, Mike thought as he pulled on his jeans and a T-shirt. The brightness of the colours reminded him of the Rainbow he'd seen the day before. He grabbed his green bomber jacket and charged out of the front door, clipping a Walkman to his belt.

During dinner the night before, Mike's mother and brother had simply brushed off his story. Mike had gone to his room frustrated, deciding to let it drop. But this morning he was excited again. He had to tell someone else about the Rainbow he'd seen. He couldn't wait to see Tissy and Pete. They were his two best friends and *they'd* listen to him.

Mike was in such a rush to get out the door, he almost tripped over a neighbour's dog on the second floor landing. "Morning, Mrs Williams," he said politely as he untangled himself from the dog lead.

"Morning, Michael," she said with a warm smile. Mike liked Mrs Williams, even though she was very prim and proper, especially about the way she dressed.

Mike flew down the stairs and passed Old Stan, who was standing outside the apartment house leafing through a newspaper. A baseball cap hid most of his thin grey hair.

"I was checking out your bike last night, Mikey," said Old Stan. "We'll have to do some work on the brakes sometime soon."

"Sure. Thanks Stan, see you later." Mike bounded down the front porch and charged around into the alley.

At the rear of the apartment building, Mike unlocked his bike and jumped on in a single fluid move. He flicked on his Walkman just as the radio announced the time was eight o'clock. Then the music kicked in, blaring from the headphones that hung loosely around his neck.

"Ignition . . . blast off!" Mike yelled. He sped down the alleyway, around the corner, and out on to the street.

He took a deep breath, drinking in the clean air. The sun had wiped away all traces of yesterday's storm and the street bustled with life. His gaze paused for a moment on a window box filled with flowers. It was the first time Mike had noticed them. The colours of the petals – they seemed so bright!

Mike blinked as he realised that since he'd seen that Rainbow, everything around him seemed so much more alive. The colours were all terrific – so bold and clear.

Mike pedalled past the gleaming silver Galaxie Diner, owned by the parents of his friend Pete. He slowed in front of Bob's Comics. The red-haired owner waved to Mike as he pedalled by. But Mike didn't stop to chat as he usually did. He was in too much of a hurry to get to school.

"Just wait till Tissy and Pete hear about this!"

Mike heard a hissing sound as he entered the science lab at Henry Hudson Elementary School. Class hadn't started yet, and his friend Pete was standing next to a large tank of helium gas at the front of the room.

Mike could tell his friend was up to something.

"Hi, Mike!" Pete said in a high, squeaky voice. He had just taken a mouthful of helium. Some of the students giggled, and Pete grinned.

A split-second later, Mr Cohen, their science teacher, came through the door. Mike and Pete rushed to their seats next to Tissy, who was cleaning her glasses with the edge of her T-shirt.

Mr Cohen noticed the helium tank was still hissing slightly and tightened the valve. "I see someone was reviewing last week's experiment with helium gas," he said with a smile.

Pete kept his lips firmly shut. One word out of his mouth, and Mr Cohen would know who had been tampering with the tank.

"However, today we're going to talk about light and colour," Mr Cohen said. He walked to the windows and started to pull down the blinds.

"But colour doesn't exist, Mr Cohen." Tissy announced as she pushed her mousy brown hair behind one ear. Tissy's parents worked for a research lab and she was constantly taking books out of the library.

Sometimes Mike thought Tissy knew more science than their teacher.

"That's right, Tissy. Colour is the sensation we get when light of different wavelengths strikes our eyes." Mr Cohen pointed a long ruler at a colourful poster on the wall. "This tiny part of the electromagnetic spectrum is called visible light. It's tiny in comparison with the energy of the universe. But don't mistake tiny for unimportant. Light is vital for life."

Mr Cohen yanked the last blind down; the room was now much darker. He took a small knife from his pocket and cut a little hole in the blind.

"Hey," Pete whispered, "our teacher is vandalising school property." Mike tried to keep from giggling out loud.

A beam of sunlight cut through the blind and into the darkness. As a shaft of gold shot across the front desk, Mike sat upright.

"Through photosynthesis," continued Mr Cohen, "light enables trees and plants to convert carbon dioxide and minerals into sugars and create the energy they need to grow. And as a by-product they produce oxygen – which of course *we* need."

The teacher pushed a large, triangular glass prism into the path of the beam and placed a piece of white card just beyond it. A beautiful little rainbow appeared on the card.

Mike felt a surge of excitement. He stared at the gleaming bands of colour.

"People throughout history have done this experiment. But it was Isaac Newton who realised it was not the prism, but the light that made the colours . . ." Mr Cohen placed another prism between the first prism and the card. The little rainbow disappeared, replaced by pure white light.

The teacher continued, ". . . and in order for plants to grow, all the energy in the different colours of light needs to be present. It it weren't, life on this planet would cease to exist."

Steve Bailey took his place with four other junior high school boys at the starting line of the track. The coach held a stopwatch, his finger hovering over the start button.

But Steve's mind wasn't on racing. He kept glancing up at a group of older kids who were sitting on the plank-seats close by.

The older group looked cool, like a group that his dare-devil dad might have joined. The boys wore leather jackets and jeans, the girls wore miniskirts and cropped T-shirts. And they all wore something that showed they belonged to the same gang – the Tigers.

Steve wished he could be a Tiger. He gazed at the brightly painted tiger on the back of one gang member's leather jacket. Then the Tiger turned round and Steve saw the look of menace on his chiselled face.

Steve shivered and shifted his gaze – to one of the girls sitting with the Tigers. She had long brown hair that shimmered across her shoulders. Her brown almond-shaped eyes were on Steve now. She gave him a long cool stare with just a hint of a smile.

He saw her round school all the time but the only thing he'd managed to find out was her name – Bridget. Steve knew Bridget hung out with the Tigers. He didn't stand a chance of getting a date with her.

Not unless he became a Tiger too.

"On your marks . . . get set . . ." The coach thumbed the stopwatch. "GO!"

Four runners bolted from the starting blocks and sprinted down the track.

Steve didn't move. He watched Bridget as she went to stand behind the Tiger with the leather jacket. Her hips swayed rhythmically to the beat of the music playing loudly on a large boom box.

The coach rushed over to him. "What's the matter with you, Steve?" he demanded. "Did you go deaf overnight?"

Then he followed Steve's gaze to where the Tigers were goofing around on the seats and shook his head sadly. "Steve, you don't want to be like those losers. You're my fastest runner, you could go all the way."

"Yeah? All the way to where?" Steve said angrily as he walked away from the track. Behind him the Tigers clapped in mock applause.

"What am I looking for anyway?" Pete asked.

"Evidence," answered Mike.

It was after school, and Mike, Pete, and Tissy were all scouting around the abandoned railroad yard.

At lunch time Mike had tried to tell his friends about the Rainbow, but he couldn't find the words to describe what he had seen. So Mike decided to take them to the railroad yard. He hoped they could find something that would prove he hadn't imagined the whole thing.

"That's him!" shouted Mike when he saw a familiar shape sitting next to a wrecked railroad car. "That's the dog from yesterday!"

Mike was happy to see the mutt again. He reached into his pocket where he had stashed half a sandwich left over from lunch. He unwrapped it and gave it to the mutt who gulped it down.

"Hey, look at that!" Tissy suddenly yelled running off. Mike and Pete and the dog followed her to the edge of a triangular crater at least thirty feet wide.

"Wow! It's a perfect triangle!" Tissy bent down and picked up some of the dirt, letting it slide through her fingers. It shimmered slightly as it fell.

"Can you test it?" Mike asked. Tissy had dozens of science kits at home, and Mike had a feeling that maybe this was the proof he was looking for.

"I don't know," said Tissy picking up another handful.

"Please?" Mike begged.

"It's only dirt." Tissy looked at him, puzzled. "What was so special about this thing you saw anyway?" she asked. "You said it came from the sky?"

"Like a UFO or something?" Pete asked. "I saw a special on TV about – "

"It was a Rainbow!" cried Mike, trying to keep from

getting as frustrated with his friends as he had with his mother and brother. "I told you, I saw a Rainbow *land*!"

Mike two friends stared at him as if he'd gone crackers.

"Let's go, Tissy," Pete said, shaking his head. "I'm supposed to be helping out at my parents' diner this afternoon."

Tissy followed Pete, leaving Mike and the dog alone at the edge of the crater. With a heavy sigh, Mike trailed along behind his two friends. The mutt sat for a moment, then slowly followed too.

3

Pete's dad was working on the counter when the three friends got to the Galaxie Diner.

"Sorry I'm late, Dad," Pete apologised. He took the coffee pot off the burner and began refilling the cups for people sitting at the booths.

Mike looked at the leftover food on some of the abandoned plates. He tried to figure out a way to sneak some of it to the mutt, who looked pretty hungry.

Pete's mother was chatting with a businessman while checking through the receipts. She'd been teaching the man a few words of Spanish with every visit.

"*Yo no te ofrezco riqueza, te ofrezco mi –* " tried the businessman.

"No, no. Pronounce it *Te of-rez-co*," corrected Pete's mom.

"Thanks," the businessman said as Pete poured hot coffee into his mug. "Or I should say, *Gracias.*"

"How's it going?" Pete asked.

"Your mother's real strict," the businessman responded.

"Tell me about it!" Pete said, sneaking a cookie off the nearby plate.

"I saw that, Peter," his mother said without even looking up.

Mike and Tissy sat at the counter. Mike was slipping part of a leftover hamburger to the mutt when Pete joined them. "Hey, maybe you saw a meteor crash," said Pete.

"Yeah? So where'd it go, Einstein?" snapped Mike, losing his patience. "Did it get bored and decide to fly off someplace else?"

Tissy rolled her eyes and Mike stormed off. Mike wasn't too surprised when his brother and mother didn't listen to him, but he never thought his friends wouldn't pay attention. The mutt followed him out of the diner.

"Hey wait, Mike," Tissy called out to him. "Don't be mad. I'll test the soil."

Steve didn't notice his younger brother following him down the street. He had only one thing on his mind – Bridget.

Bridget . . . Bridget . . . Bridget . . .

Steve said the girl's name to himself over and over at night before he went to sleep and in the morning when he woke up. He dreamed of her riding behind him on a motorcycle. Riding out of town, into the sunset. He could almost feel her arms around his waist as they leaned into the turns.

Steve knew where Bridget lived and where she hung out. Maybe if he wandered around her neighbourhood long enough, he'd find her.

A girl with long brown hair was walking up ahead. He admired her graceful legs flexing beneath a miniskirt as she walked. She turned once to look in a shop window and Steve saw her face.

Bridget.

She looked even more beautiful than she had yesterday at the sports field.

Steve walked towards her. He nervously mouthed the words he wanted to say, practising for the moment when he would finally summon up the courage to speak. But before he could say anything, she abruptly turned in through an archway. Steve took a deep breath and followed. This might be his big chance, he thought. If he could talk to Bridget alone, he could ask her out.

When Steve swung into the open archway, Bridget stood before him, leaning provocatively against the wall, her long brown hair glistening in the brilliant sunlight. But they weren't alone. Behind her, in the courtyard, stood the Tigers gang.

"Hey, Bailey. Whatcha doin' here?" one of the gang shouted.

"Shouldn't you be running in circles round the track?" a second taunted.

"Hey, look! Bailey brought his kid brother for protection," one of them called.

Steve glanced behind him and saw Mike standing in the archway.

"What are you doing here?" he said angrily.

"I followed you," Mike said. "I had to tell you some – "

"Kid!" one of the Tigers interrupted. "You'd better

get your big brother outta here before we rearrange his face."

Of all the times for Mikey to follow him, he thought angrily. Now he was totally humiliated in front of Bridget and the Tigers. Steve turned around to leave. There was no point in staying. Nothing was working out the way he wanted.

"Come back and visit another time, Runner-boy," a Tiger taunted. "Then we'll see how fast you really are."

The other Tigers laughed, but it wasn't a friendly laugh. It sounded like trouble.

"Don't you ever embarrass me like that again," Steve hollered at his younger brother.

Mike had to run to keep up with Steve. He knew his brother was angry with him, but it wasn't his fault. How was he supposed to know where Steve was going?

Mike decided not to say anything until after Steve had calmed down. Now definitely wasn't the time to tell him about the Rainbow.

The brothers walked in silence past a pawn shop with bold letters declaring GOLD: BOUGHT & SOLD. They made their way towards the Motorcycle Bonehouse nearby.

A group of burly-looking Hell's Angels sat sharing beers. The Angels were surrounded by motorbikes old and new. A couple of the bikers nodded hello as the boys entered. They all knew Steve by sight, because he was always in there admiring the bikes.

Steve ran his fingers across the blood-red leather and

silver detailing of a gleaming black Triumph motorcycle. "It's hot isn't it?" said Steve.

Mike moved over to Steve and pretended to show an interest in the bike. This was the perfect time to tell him about the Rainbow.

"What if I told you something really hot?" asked Mike. "Something nobody'd believe? Well, no one except maybe Gramps."

"Gramps! Get real, Mikey." Steve continued to stare longingly at the Triumph. "Now *this* is real. A primo piece of machinery! Imagine riding out of here, away from this crummy town. Just like Dad did."

Mike was getting frustrated. This wasn't going the way he had hoped.

"This was his kind of bike," Steve went on. "Like the one he used to ride through the Wall of Fire, blasting through the flames, crowds cheering him on." Steve motioned for Mike to get on, then he straddled the seat behind him. "When we were with the carnival, before you were born, Dad used to take me out on his bike, sit me up front like this." He leaned forward almost hugging his little brother.

Mike tried to imagine what it would have been like to have known his father the way Steve had. To have his father sitting behind him on the bike.

"Man, we used to go fast," said Steve. "I felt like the coolest kid in the world. That's what it's all about Mikey. Freedom."

"But freedom don't come cheap, Stevie, my boy," said one of the mechanics, a big guy with a heavy moustache and leathers. He was fixing one of the half-dozen broken bikes nearby and had overheard them.

Another biker watched the mechanic as he worked and shouted over at him. "Seems a little young for that bike. Don't look like he could handle that much power, hey, Butch?!"

The bikers laughed.

Steve looked as if he'd been stung.

Mike was suddenly desperate to think of a way to restore the connection he'd felt with Steve a moment before. In fact, Mike was so distracted that he didn't even notice the strange mutt watching them intensely from the doorway.

Later that afternoon Mike's mom was rushing around the apartment like a whirlwind.

"Sorry sweet pea, I can't talk now," she said to Mike, grabbing floppy disks and books off the table and stuffing them into her bag.

"I've got to get to the office for an important meeting. If I can sell Keefe on this segment he may give me a shot at something other than 'Jersey Hijinks.' "

Mike knew his mother was unhappy with her job working on local news stories. She wanted to get back into investigative reporting. She had been a successful journalist before she had met his father at the carnival. But after they got married, she'd given up her career to raise her family.

Mike was disappointed. He'd already waited so long

to tell her about the Rainbow. "Okay, I'll go and visit Gramps."

Jackie tensed. "Well, be sure and eat first." She grabbed her coat and started to put it on as she moved towards the kitchen. "Lasagna's in the freezer. Microwave it on – " She stopped abruptly in the doorway. "Michael. What's that dog doing in our kitchen?"

Mike looked through the doorway. The mutt was sitting there, his tail wagging happily.

Mike was pleased to see the mutt. Ever since the dog had led him to the Rainbow, Mike had felt there was a bond between them. He was about to explain about the mutt when a car started honking loudly outside. It was Jackie's boss, Keefe Russell, waiting to give her a lift.

"I've gotta go. Make sure that dog's out by the time I get back," Jackie said grabbing her coat.

"Bye, Mom," Mike said.

The mutt barked his goodbye too, and Jackie disappeared out the door.

Mike stood in his grandfather's living room holding a framed photo of Frank Bailey back in his carnival days. In the photo Gramps was sitting outside a trailer with a young Steve on his lap. Standing next to them both was a much younger-looking Jackie and beside her stood Mike's father; holding a baby.

That's me, thought Mike, looking at how happy they all seemed. A happy family.

Frank's voice made him jump. "Where did you find

your friend?" he asked placing a bowl of food on the kitchen floor for the mutt.

"Actually, he kinda found me," Mike explained.

Frank bent down and stroked the dog. "He got a name?"

"No . . . er . . . yes . . . it's Mutt!" Mike announced. Mutt's ears perked up. Mike was pretty sure the dog liked his new name.

Frank gently took the photo from his grandson's hand and put it back down on the table. "Crazy days, great times. Your mom was beautiful, huh. Still is." He smiled at the picture. "We had fun, travelled the whole country. Seemed like there wasn't anything we couldn't do if we wanted."

"Yeah, then I was born," Mike said. "One kid is easy to deal with, but two – I guess I ruined everything."

"C'mon, it wasn't like that and you know it," Frank said as he led Mike over to the dining-room table. He returned to the kitchen and brought out two slices of pizza on paper plates.

"It was just time to settle down," Frank explained. "Can't run forever. Life changes, and when you've got responsibilities, you've got to take things a bit more seriously. It's a sacrifice that's worth it, and your parents were thrilled when you were born."

"So why did he leave us then?" Mike asked.

"Sometimes things happen . . . We all make mistakes," Frank replied.

But Mike had heard that answer before, and it just didn't sound like much of one to him.

Picking a piece of pepperoni off his slice, Mike placed it in his mouth. "It's just . . ." he began as he chewed, "I wish . . . maybe you, Mom, me and Steve . . . we could get together sometimes, you know, like a family . . ."

A moment passed and Frank looked thoughtfully at Mike. "You got something to tell me?" Frank asked his grandson.

Mike took a deep breath. Sometimes it felt as if his grandfather could read his mind. Mike wanted to tell him about the Rainbow. But so far, no one had believed him. What if his grandfather laughed?

"I saw a Rainbow," Mike began. "Not just any rainbow . . . it was, like a magic Rainbow. I saw it land . . . right in front of me." He waited for a reaction from his grandfather, but there was none. Only silence. "You don't believe me either." Mike stared down at his slice of pizza. He didn't have an appetite anymore.

"What do you mean?" Frank's voice was filled with excitement and his eyes gleamed. "Of course I believe you! You saw something most people only dream about – real magic!"

Mike's eyes were suddenly as bright as his grandfather's. "Do you think I'll ever see it again?".

Frank gripped Mike's hands. "See it again? Don't think it's going to come to you. You've got to go out and chase that Rainbow. Mikey, don't dream the dream, *be* the dream."

That's it! realised Mike. In that instant, he knew what

he had to do. And he also knew he'd need the help of his friends.

4

Pete watched as Tissy scooped soil from a test tube into a Petri dish. She had set up a mini science lab at one of the booths in the diner. She peered through a magnifying glass at the soil sample she had collected from the crater in the railroad yard.

Suddenly her eyes widened.

"I wonder . . .?" she murmured, then she dripped a few drops of water onto the earth.

The moisture soaked in and suddenly bright gold beams leapt out at them. The brilliant rays of light reflected off the chrome, illuminating the entire booth. Then disappeared with a whoosh!

"Awesome!" Pete said, wide-eyed.

"Double awesome!" agreed Tissy.

"Maybe that's what Mike saw at the railroad yard," Pete said, still staring at the tabletop in case there was more to see.

Just then, the door of the diner swung open and Mike came in with Mutt at his heels.

Pete and Tissy jumped up to meet him, their words tumbling over each other in their excitement.

"We want to find the Rainbow, too!" they cried, grabbing hold of him.

Mike looked from one to the other in astonishment. "Great. But how do we find it?"

"We can use a theodolite," Tissy announced.

"What's a theodolite?" Mike asked, confused.

Tissy explained that a theodolite was a device for measuring angles. She knew of a way they could use it to help them collect data.

"We'll also need a computer program to interpret the data from the theodolite and tell us exactly where the Rainbow landed," Tissy continued.

"And our bikes need to go faster so we can get there before the Rainbow disappears," Pete added.

"Whoa!" Mike said. "I'd better make a list."

He grabbed a napkin off one of the tables and took a pencil from next to the phone. Then he began scribbling out his list:

STUFF TO DO
Make theodolite
Write computer program
Fix bikes

The three friends divided up the tasks. Over the next week, Tissy would write the computer program, Mike would be in charge of the bikes, and Pete would build the theodolite.

After school every day, Pete set up one of the diner booths as a work table to make the theodolite. He gathered odd bits of wood and other stuff he found in the tool shed out at the back. Working from a diagram

Tissy had drawn for him, he made these into theodolite parts.

As the week went by several of the regular customers grew curious and began to help Pete build the device. Finally, after a lot of hard work, Pete fitted the telescope to the main body of the theodolite. Now all he had to do was find a tripod to stand it on. One of the waitresses said she had one she could loan him, and she rushed home to get it.

In the meantime, Tissy was using a computer at school to design the computer program they needed. She was the first student in the school computer room in the mornings and the last one to leave in the afternoon. Mr Cohen was very impressed with her dedication to her schoolwork as he peered through the classroom doorway every day to see her sitting in front of the screen.

Old Stan helped Mike tune up the bikes. Together they tightened the chains, pumped up tyres, and checked the brakes and gears. Even Mrs Williams helped out, careful not to get grease on her good clothes.

It was a long, hard week. But by the end, Pete, Tissy, and Mike had finished. Mike crossed the final item off their list.

They were ready.

Mike tried to look attentive as Mrs Burns began her geography lesson. "Okay, students, the last time we were together we talked about tropical rainforests and

"Ladies and gentlemen, observe!" announces Frank Bailey as he begins his magic show.

Mike Bailey listens as his teacher,
Mr Cohen, explains to the science class
the importance of colour and light.

Mike, Pete, and Tissy examine the crater at the railroad yard where the Rainbow touched the ground . . . and Mutt makes some new friends!

When the Rainbow touches down, Mutt wants to be the first one to ride it.

Mike, Pete, Tissy and Steve follow Mutt into the glorious colours of the Rainbow.

Awesome! Fantastic! Spectacular! Tissy can't believe she's actually riding a Rainbow!

As the kids fly through the Rainbow, they join hands and spin like a top in the dazzling colours.

The kids land in Kansas – and in the office of an unhappy Sheriff Wyatt Hampton!

As the world's colours slowly fade away, everything goes haywire! Riots break out in the streets.

"Go, Mike, go!" Frank, Jackie, Mr Cohen, Tissy and Pete watch from a rooftop as Mike rides the Rainbow again and tries to return its missing gold.

Mike did it! He's healed the Rainbow, and the colours of the world have finally returned.

their importance to our planet. Of course, this is just one of the vital ecosystems . . ."

Mike tapped his pencil impatiently against his leg as his gaze strayed out the window. He and his friends had been ready for nearly two weeks now, but the weather was not co-operating. There hadn't been so much as a drizzle.

This morning promised yet another sunny day. Mike stared at the bright blue sky – not a very promising condition for the Rainbow.

Sighing with disappointment, Mike turned his attention back to his teacher. He hadn't noticed the almost invisible flecks of rain that began to fall, brushing like tiny tears against the glass.

". . . the other three are the Laurentian forests, the Marine Ecosystems of the St Lawrence, and the Polar World . . ." Mrs Burns continued.

Woof! Woof!

Loud barking suddenly interrupted Mrs Burn's lecture. Mike recognised the sound and raced across the room to the window. The sprinkle had built to a steady rainfall and Mutt stood below, hackles up and nose raised high as he barked at the sky.

Mike followed the dog's gaze and did a double take. Materialising in the sky were the glorious colours of the Rainbow!

"What's the matter, Michael?" asked the teacher.

Mike turned towards Pete, face glowing, eyes burning. "It's here!"

"Michael! Michael? Please sit down, Michael," Mrs Burns repeated.

But Mike and Pete were gone, running from the classroom, leaving Mrs Burns speechless.

As the two boys raced into the empty corridor, Tissy burst from an adjacent classroom.

"The computer room, ninety seconds," Mike shouted at her on his way to his locker.

"I'll get the theodolite!" Pete called as the three of them took off.

Mike grabbed his camera from his locker and then ran towards Pete's locker.

Tissy raced off in the opposite direction, passing Mr Cohen who had just rounded a corner. Luckily, the science teacher was busy leafing through papers and Tissy managed to skid round the corner before he even looked up.

Meanwhile, Pete handed Mike the theodolite.

"Get our helmets and meet us at the bike rack," said Mike. Pete nodded and headed off as Mike turned towards the exit to the roof.

Mike climbed out onto the school rooftop and gasped when he saw the band of beautiful colours that reached from the sky and swept down over the city. For a long moment he was spellbound, then he got to work. Quickly he set up the theodolite and took the readings they needed to locate where the Rainbow had landed.

Tissy's fingers flew across the keyboard as she hurried

to access the program. "C'mon. C'mon, c'mon!" she mouthed to herself.

Mr Cohen was still in the corridor when Mike rushed past, carrying his notebook. This time, the teacher looked up from his papers. Puzzled, he wrinkled his brow for a moment, then turned to stroll in the direction of Mike's speeding body.

The second Mike reached Tissy's side he flipped open his notebook which was a little soggy from the rain, and began reading. "West twenty degrees. South thirty-five. Distance one-point-five miles."

Tissy tapped the keys of the computer and the unit buzzed and whirred as if the data she'd fed it were tasty morsels it couldn't wait to chew and digest. Next the monitor came alive with electronic pulses that shot across the screen. A map of Hudson Harbour appeared and the cursor skimmed across the city, tracing a path down to the river. The cursor hit its target and the printer jumped to life, spewing out a map.

Mike ripped off the piece of printer paper. He and Tissy burst through the door – just as Mr Cohen was about to come in.

"What the – "

But Mike and Tissy weren't slowing down now. They left the teacher spinning, his papers flying, as they fled past him and down the stairs.

Pete had already unlocked two of the bikes and was snapping open the lock of the third. He wrenched the chain free just as his friends arrived out of breath. The three leapt into the saddles like cowboys and were fastening their helmets when Mike's brother ran over to them from the junior high school bike racks across the street.

"Steve?" Mike said. "What are you doing here?"

"I could ask you the same question, Mikey," Steve said. "What are *you* doing?"

"Chasing that," Mike shouted, pointing.

Steve looked to where his brother pointed high in the sky. It was the Rainbow – a translucent arch of colours through silvery drops of rain.

"Oh, well. I was about to cut my maths class anyway!" Steve yelled and ran back to get his own bike.

"Let's go," Mike shouted and they were off, speeding through the schoolyard and out of the gates.

Tissy, Steve, and Pete flew along the sidewalk with Mike up ahead and Mutt speeding alongside like a bullet. They rounded a corner and turned right, wheels splashing through puddles as the rain continued to fall in a hazy mist.

The rain had almost stopped and the sun shone even brighter as they accelerated along the canal path.

"C'mon. We're running out of time!" Mike yelled as he looked up.

Blurred wheels spun as they pumped the pedals faster and faster. The sky looked as though it were clearing and the sunshine washed over the ground behind them as they sped further and further away from the crowded landscape of Hudson Harbour.

They whipped across another road and into the city's deserted docks, charging under huge iron cranes and girders, past the towering grey walls of abandoned factories.

Mutt darted ahead barking, leading the gang forward.

"We have to go faster! It's disappearing," Mike warned.

They reached the dock perimeter, bumping over uneven ground as they raced time and the Rainbow.

"That's it! Around there!" Mike called, urging everyone on.

Fifty yards, forty yards, thirty – with Mutt bounding forward, the group skidded round the corner in formation and braked to a stop at a wooden fence. All five gazes lifted skyward. Four kids and a dog were instantly mesmerised.

Mike looked up, beaming, eyes full of wonder. Stretching down from the sky was the enormous, pulsating Rainbow! It was a glorious arc reaching into the heavens and disappearing into the highest clouds. To Mike, the magnificent bow of gleaming colours seemed even more beautiful than when he had seen it the first time!

"Amazing!" Steve murmured.

The others were speechless.

Pete, Mike, Tissy, and Steve dismounted and leaned their bikes up against the fence, then hauled themselves over the top to get closer to the pulsating colours.

Mutt had already scampered through the fence's widely spaced slats and was now sitting near the foot of the Rainbow. The dog was bathed in brilliant colours, which spun and danced in glistening currents. The children stared in amazement as they stepped closer and closer.

It was beautiful . . . remarkable . . . extraordinary!

Mutt darted forward and jumped into the Rainbow. Electrical charges seemed to reach out and engulf him, leaving an outline of glittering particles in his place. Multicoloured sparks leapt from the Rainbow and swirled round the children's bodies. Tiny haloes of colour formed round their feet.

Then suddenly, the whole group was yanked down and swallowed by the earth, disappearing in a balloon of dust! Tissy's helmet hovered for a moment in midair – and then it dropped to the ground.

Before Mike had time to figure out what was happening to them, he saw that they were all now *inside* the Rainbow.

He couldn't believe it. They were really inside it.

Then gently, the Rainbow lifted them. It was pulling them up, into the air.

5

Mike was completely amazed. He felt so light, as if he could fly. And he *was* flying! He glanced around at his friends. They were all flying too, and they all seemed stunned!

He looked up, into the Rainbow. As far as he could see, bands of brilliant light and dazzling colours stretched before him. Mutt had already glided far, far ahead, and the Rainbow seemed to go on for eternity.

The fantastic hues spun and cascaded all around. Mike reached out, waving his hand through the web of light, creating colour trails that shimmered from his hands and every moving part of his body. His friends were also illuminated by the many changing colours. Reds, oranges, greens, and purples trailed from their bodies in shimmering streams.

Suddenly, with a rush of air, the Rainbow propelled the group further along. They whizzed by the Statue of Liberty and the Manhattan skyline quickly began to shrink beneath them.

As they moved higher and higher, Mike gasped at the glorious countryside. It was an incredible vista of green meadows, clear blue lakes and jagged purple mountains.

It was totally awesome. Amazing, unbelievable. Mike felt so free. He reached out his hand to Pete. Pete grabbed hold, then reached out to Steve. Steve hung on and then joined hands with Tissy. Finally, she caught Mike's hand to complete a human circle.

The four began to spin like a top. Around and around they went, laughing with giddy excitement. Colour trails spun from their bodies, light bouncing off every moving part of them. Then they let go of each other and spun away.

Pictures, Mike reminded himself. I've got to get this on film. Mike struggled to get his camera from his pocket. He began clicking off shots in every direction – snapping his friends, the landscape far below, and the Rainbow.

Gently they floated ahead. They seemed to be entering a new part of the Rainbow. Here the clear colours were giving way to the brightest, whitest, purest of light.

The light soon transformed itself into a glistening sea of golden sparks that twinkled and glowed like stars.

Steve floated a little way off by himself, slowly, trailing his hands. He ran his fingers through the golden streams, then clasped his hand tight. Bringing it up close to his face, he slowly unfurled his fingers. He couldn't believe his eyes. The gold had solidified and formed a perfect golden nugget in the palm of his hand.

Steve stared at it for a moment and then quickly slipped it into his jacket pocket. Reaching out again, he grabbed at the golden streams. As the second nugget formed, a light crackled around him and there was a gentle rumble. Golden tendrils shot out, trying to connect with the other golden particles and reform the pattern, but the connection had been broken.

Mike, Tissy, and Pete looked up, startled. Something had happened, Mike could feel it.

Steve reached out a third time, clutching at the Rainbow, grabbing a third piece of gold. There was a blinding white flash and a pounding of thunder as the children were thrown apart. Suddenly the children and Mutt were violently spat out into midair!

No longer carried by the Rainbow, they plummeted at an incredible speed towards the earth. They shot through clouds, the landscape below becoming more and more distinct as they got closer. Mike's heart was pounding so hard, he thought it would burst out of his chest. He couldn't see his friends or his brother, and there wasn't time to look or think as the ground came rushing up towards him.

They were seconds from a brutal crash-landing, when the Rainbow suddenly reached out and snatched them up. Once again they were being cradled by the Rainbow, only now they were suspended just a few feet above the ground.

Slowly, gently, the Rainbow lowered them to the ground. Then the brilliant sparkling band of colours shot up into the sky and was gone.

Dazed, Mike blinked and tried to look around him. His eyes seemed to need time to adjust and focus. He rubbed them and blinked again.

The colour green appeared before him.

"I'm in a field," he said aloud. He realised that he

was standing under an enormous canopy of dense green leaves. Tall stalks towered over him, at least ten feet high. He looked up to see plump cobs of corn and clouds floating along in a sky of clear blue.

"Mike, Tissy. Is anybody there?" The voice was Pete's. "Where are we?"

"Pete! Over here," Mike answered.

"Mike!" shouted Tissy.

"Mike, Tissy," Pete echoed.

For several minutes, the friends wandered in and out of the tall stalks of corn, narrowly missing each other.

Then Steve's voice came out of nowhere.

"Mike! Mike, where are you?" Steve shouted, soon running smack into his brother. "You okay?" he asked, gripping Mike's shoulders and hugging him tightly to him.

"Sure," said Mike, his relieved voice muffled against Steve's T-shirt.

The two brothers began to search together. After several more minutes, they found Tissy and Pete.

"I don't believe what happened – and I was there!" Tissy exclaimed.

"Yeah. But I've got proof. Pictures," Mike said taking the camera from his pocket.

"That last bit was pretty scary," Pete said. His skin seemed to have taken on the colour of their surroundings.

"You feel okay?" asked Mike.

"Yeah," said Steve. "You look kinda green."

Pete nodded. "I'm a little shaky, but I'm not gonna be sick or anything."

Woof! Woof!

It was Mutt. He began running ahead and then coming back for them. Thinking the dog was trying to show them the way, Mike followed Mutt through the corn, the others close behind.

They continued to talk as they headed off in single file through the rows and rows of corn.

"Mike, where are we? And where're we going?" Steve asked after they had walked for a little while.

Mike shrugged. He had no idea. But the sun was low in the sky and getting lower. Soon it would be dark.

"I think we're going in circles," said Pete, after they had wandered for what seemed like hours.

"Let's be logical about this," Tissy suggested.

"Logical! We just found a Rainbow, rode it through the sky, and ended up heaven knows where. Which part should we be logical about?" Pete asked sarcastically.

"It could have been just a mirage. Maybe it was caused by shock," Tissy went on.

"Well, this is real," Mike said breaking off a piece of the stalk in front of him and tapping Tissy with it.

Woof! Woof!

Mutt seemed to agree with Mike.

Steve stopped a moment and secretly looked into his pocket. The gold was still there. He smiled. It was *real* all right.

"Well, which way should we go?" asked Pete.

Steve abruptly stopped. "I've got an idea. Tissy, get on my shoulders."

With Mike's help, Tissy managed to scramble up onto Steve's shoulders. Now she could see over the corn.

"Over there!" Tissy said, pointing. "I see some farm buildings!"

They walked in the direction of the farm. Soon they were out of the cornfield and walking through a lush green garden behind a whitewashed house. A battered old station wagon was parked in the shade of an old apple tree.

Mike led the way, and as the four got closer, they heard whistling. It was coming from the apple tree. They looked up, expecting to see some sort of bird, but instead they saw a little girl whistling at the sky. She was about six years old with long blond hair and freckles. She was dressed in hand-me-down jeans and a flowery yellow shirt.

"Hi there," Mike said. "We're kind of lost – "

"And we've been walking for ages," Pete added.

"Is anyone home? Your mom and dad, maybe?" Mike asked.

She didn't answer.

"Listen, kid. Is there anyone around here or not?" Steve demanded.

The little girl looked at each of them then spoke in a small voice, "You fell down from the sky."

Mike was astonished. "You saw us?"

"I saw colours," she replied.

The children looked at one another.

"I was waiting here for you. I knew you'd come and say hello," she said.

A short time later, the four kids and Mutt were crammed into the back of the old station wagon zooming off down the dirt road and throwing up dust in their wake.

The girl's father wasn't very talkative. When he did speak, he had such a thick accent, the kids had trouble understanding him.

"Can you tell us again, where we are?" Tissy said.

"Nearest town's Sativa Falls," the farmer replied. "That's where we're headed. The sheriff 'll know what to do with y'all, cause I sure as thunder don't."

Sativa Falls? Mike didn't know any town by that name. "We *are* in America . . . aren't we?" he asked hesitantly.

"O' course y'are, son. You're in the smilin' state o' Kansas."

"Kansas!" Mike, Tissy, Steve, and Pete all exclaimed together. After that, they couldn't think of a single thing to say. They were stunned into silence.

For the rest of the ride, they quietly watched the sun slowly slip down to the horizon. Vast fields of corn spread before them as far as the eye could see.

6

Mike had never been inside a police station before. Tissy, Pete and Steve seemed as nervous as he felt. No one had said a word since the deputy had told them where to sit to wait for the sheriff to return.

A secretary sat at an old wooden desk, snapping chewing gum. Her fingers banged on the keyboard of an ancient typewriter.

The deputy opened a round tin of tobacco and placed a huge chew in his mouth. A few moments later he spat dark tobacco juice at a bucket on the floor – and missed. Totally.

Mike stared at a plaque on the wall. "Sheriff Wyatt Hampton," read Mike aloud. Just then the door crashed open and a large man strode in wearing an immaculate blue uniform and dark glasses.

"Okay, Clarence," said the large man. "What's all the excitement about?"

"Well, sheriff," began the deputy as he motioned towards the children. "It's these kids here . . ."

The sheriff 's eyes travelled down the row from Tissy, to Mike, to Pete, to Steve, to Mutt – whose ears pricked upright, tail wagging.

"I don't like kids," the sheriff muttered as he moved

towards his private office. "And that's the strangest lookin' dog I've ever seen."

Mike and the others followed the sheriff inside his private office. The sheriff sat down and leaned back in his chair with his hands behind his neck. The kids told him what had happened. But it was obvious he didn't believe a word they said.

"A rainbow? And you rode it through the clouds over a thousand miles? Sure you did, kid!"

"But it's true," Mike insisted.

"And I s'pose you found a pot of gold at the end of it, right?" the sheriff added sarcastically.

Steve looked away sheepishly while the other boys rolled their eyes. Only Tissy smiled.

Seeing Tissy smile must have given the sheriff a new idea. He changed his tactics. Instead of trying to bully the story out of them, he tried to charm it out.

He came round his desk towards Tissy, careful to avoid Mutt. The dog was growling at the sheriff, who continued to stare at the man as if he were a juicy steak.

The sheriff knelt down next to her chair and pasted a smile onto his face. "Patricia, isn't it? I want you to trust me, child. I am your friend, Sheriff Hampton. Now, Patricia, I want you to remember all the way back to where you woke up this bright and sunny morning."

Tissy sighed and then spoke in a very adult voice. "I know I'm just a child, but that doesn't mean you have to treat me like one." She glared at him as her friends sniggered.

Tissy's remark caught the sheriff off guard. He

quickly went back round to his chair, keeping as far away from Mutt as possible by squashing himself up against the desk. "What is that thing anyway? It looks like its parents were a raccoon and a wolverine."

"When can we go home?" Pete pleaded.

"Where is your home, sonny?" the sheriff asked.

"Here we go again," Steve said exasperated.

Thwut! The deputy spat again.

"Can we just use the phone?" Mike asked.

Back in New Jersey, Jackie Bailey was sitting in the TV news room with her boss, Keefe Russell, a beady-eyed man who wore his hair in a ponytail.

The place was bustling. Banks of television monitors blared, phones rang non-stop, and people constantly rushed back and forth clutching stacks of files and videotapes.

Images played on a small monitor on Jackie's desk as she showed Keefe a news segment she'd produced. When it was over she waited nervously for his reaction.

"It's a good piece, Jackie," said Keefe. "Maybe go a little lighter. Let's get Libby Evans to cover it."

Jackie's face fell with disappointment. "Libby? But this is an editorial piece – "

"Let's talk about this some more over dinner," Keefe interrupted, putting his arm round Jackie's shoulders. The shiny suit he wore reminded her of reptile skin. "How about my place at – "

"Jackie!" someone called from the other side of the news room. "Line six!"

Saved by the bell, Jackie thought as she slid from her boss's unwelcome embrace. She picked up the receiver and heard her younger son's voice trying to explain where he was.

"A police station!" Jackie's alarmed voice echoed through the news room. "What happened? Are you all right?"

Back in Kansas, Mike was sitting on the sheriff's desk as he spoke to his mother. Pete was leaning against the sheriff's chair. But the sheriff himself wasn't paying attention to the kids. He sat on the edge of his desk flipping through a magazine.

"Uh . . . I think you'd better tell her . . ." Mike said, holding the phone receiver out to the sheriff.

The sheriff took the phone. Mike could tell he wasn't pleased with the situation, but his voice didn't show his irritation. It was the same phony sweet voice he had tried on Tissy. "Good evening ma'am. Sheriff Wyatt Hampton here. Sativa Falls County Court House, State of Kansas. And to whom do I have the honour of speaking . . .?"

Meanwhile, in a far corner of the news room, another woman was listening to the conversation on an extension. Reporter Libby Evans curiously raised an eyebrow as a stunned Jackie repeated her name to the sheriff.

When the conversation was over, Libby Evans placed the receiver down and Keefe Russell bent close to her. "I want you to put a crew together right now. Got a hunch about this one. Could be a nice little domestic story."

Jackie was already grabbing her things and about to rush out of the news room when Keefe stopped her. "It's unbelievable what those kids did," Keefe said. He paused for a moment. "You know, Jackie, it'd make a great story – "

"What! No way, Keefe. Don't even think about it." Jackie tried to get round her boss. She had a lot to do. Call Tissy and Pete's parents, arrange for the children to fly back on the next flight, meet them at the airport . . .

"C'mon, you know me better than that," Keefe responded as Jackie finally succeeded in sidestepping him and rushing out the door.

Pete, Tissy, Mike, and Steve stood in the central concourse of Wichita Airport. A sea of passengers, pushing trolleys and dragging suitcases, flooded by them.

The sheriff and his deputy stood guard over the children as if they were hardened criminals.

A baggage official came for Mutt who was now imprisoned in a dog carrier for the flight. Mike's fingers clung to the wire door after the baggage man loaded Mutt onto his cart. "Don't worry, boy," said Mike, starting to run after the cart as it rolled away. "It's only for a few hours. It'll be okay."

"Hey, get back here!" shouted the sheriff, grabbing Mike by his collar. "Now sit down," he ordered as he guided the four children to a bench.

Four pairs of eyes followed the movements of the

sheriff as he paced back and forth in front of them. Watching the sheriff was like watching a tennis match.

"Okay, your flight leaves at eight-thirty. Not soon enough as far as I'm concerned. I do not want to see any one of you move even an inch, or else." He held up a pair of handcuffs to make his point.

Then the sheriff headed for the phone, the deputy right on his heels.

Four angelic faces smiled back as the kids sat perfectly still on the bench.

The sheriff dug into his pocket to find a dime for the phone. Splat! The deputy spat noisily onto the floor.

The sheriff screwed up his face in disgust. "Clarence, will you stop doing that. You are an environmental hazard! And what are you doing here anyway? You're supposed to be watching those . . ." He swivelled around and did a double take. "Oh, no!"

The seats were empty!

7

Tissy, Pete, and Steve tore down the concourse, with Mike leading the way.

"I really don't think this is a good idea," Pete shouted.

"What are we looking for anyway, Mikey?" Tissy yelled.

"Proof," Mike shouted behind him as he scanned the shops. They ran past a gift shop, a pharmacy, a newsstand, a Fotoshop – a Fotoshop! The group did a sudden about turn and came to a stop in front of a large poster. It read: ONE HOUR PHOTO PROCESSING.

Mike looked at the others and smiled. "Bingo!"

The four kids hurried inside.

The shop was all bright yellow plastic with a large processing machine at one end and a long green counter running along the back. A dark-haired young man with a moustache whistled loudly to himself as he stacked boxes of film on a shelf.

Mike slammed the roll of film down and snapped the back of the camera shut. "We need this film developed right away. Our flight leaves in thirty-five minutes," he announced breathlessly.

The young man pointed at the sign in the window. "Can't you read? The sign says one hour, and that's how long it takes. Besides, I close up in exactly thirty-two minutes, so you can just come back tomorrow." The man turned his back on them and went back to stacking the film.

"But it's a matter of life and death!" Mike begged.

The Fotoshop man turned around very slowly. He placed his hands on the counter and pressed his face close to Mike's. "It sure *will* be a matter of life and death – yours! – if you don't stop bugging me!"

"Psst!" Tissy grabbed Mike's arm and pointed. Through the shop window they could see the sheriff head their way. As he passed close by, four heads ducked beneath the counter.

The Fotoshop man watched the sheriff pass and whispered to the kids hiding below. "What gives, you on the run or what?"

"Sort of. That's why we've got to get the pictures developed. They're our proof," Mike whispered back.

The young man's eyes suddenly lit up. "Well slap me silly!" he said ducking down to join them.

"What are you doing here?" asked Mike, "He's not chasing *you*!"

The young man suddenly jumped up. "That's right. Act natural. Don't attract attention. So, when's your flight again?"

"Thirty minutes," Mike responded hopefully.

The man grinned. "I'll have your pictures ready in twenty."

Just then, the sheriff walked by the shop and stopped right outside the entrance.

"Uh, oh," the Fotoshop man said nervously. "Sheriff 's back." He motioned for the kids to scoot around to his side of the counter so they'd be out of sight.

The sheriff strode into the Fotoshop and straight up to the counter. "Have you seen four kids running around here, son?"

The young man smiled stupidly at the sheriff. "Uh, n-no, s-sir," he stammered nervously.

Mike knew they had to get out of there fast. He didn't trust this guy to cover for them. He led the others one by one, and they crawled on their hands and knees round the side of the counter and out of the door.

Steve was the last one round the side of the counter. He'd got past the sheriff 's legs and was almost out of the door when the sheriff suddenly looked around.

"Hey!" shouted the sheriff. "Come back here you little barn weasels!" Then he was in motion, chasing Steve out of the Fotoshop.

Steve bolted in one direction, Pete took off in another, while Tissy and Mike ran straight ahead.

Tissy grabbed a large suitcase and slid it straight at the sheriff. Then she jumped over a bench, narrowly missing a mother and son waiting for their flight.

Bang! The case hit its target and the sheriff tripped and went flying! He fell to the floor, landing flat on his stomach. His big body went sliding across the floor, crashing right into a metal bench!

Two startled travellers stared down at the rumpled figure. "Excuse me," he said groggily as he picked himself up.

Then the sheriff caught sight of a figure ducking behind a pillar in front of the Country Clothing Shop. It was Tissy.

The Sheriff marched over, eyes trained on Tissy as she moved through the shop. She dodged between the racks of clothes and shelves full of gifts and cowboy boots. Then she disappeared from view.

The sheriff rushed into the shop, eyes darting every which way, just as the changing room curtain swished closed.

He moved in slowly, a huge smile spreading across his face. "Okay. I've got you now," he said as he threw open the curtain.

A woman inside was changing clothes and had nothing on but her underwear. She screamed in surprise.

"Oh m-my, I'm s-sorry," the sheriff stammered – and the woman smacked him hard across the face!

Staggering backwards, the stunned sheriff let the curtain fall back into place.

From behind another curtain, Tissy giggled silently. Then she carefully sneaked out of the shop, leaving a red-faced sheriff to explain to the angry shop manager what he was doing in the women's changing area.

Mike, Pete, and Steve had just dodged the deputy when they saw Tissy running towards the departure gates with the sheriff once again in hot pursuit.

"Okay, now the line has been crossed," shouted the sheriff. "You are breaking the law."

Out of the corner of his eye Mike spotted a baggage trolley and grabbed it from a surprised passenger. With Pete and Steve's help and a good running start, Mike gave the trolley a huge push that sent it rolling forwards at an incredible speed.

Tissy had reached the barrier and there was nowhere left for her to go. She looked panicked as two airport officials grabbed hold of her. Seeing she was trapped, the sheriff stopped running. He walked slowly towards her, a big, cocky grin on his face and his hands on his hips. He didn't see the trolley behind him, rolling straight for him!

"Well now, little lady. It's reckoning time – " just then the trolley hit him. With a thud, he fell backwards into it. The sheriff was wedged in and the trolley was going faster and faster!

"*YEEEEEEEAAAAAHHHHHHHHHH!*" His arms and legs flailed in the air, he shot the entire length of the airport concourse!

WHAM! The trolley slammed against the counter, catapulting the sheriff through the air and sending him headfirst along the conveyer belt and into the X-ray security scanning machine.

Lights flashed, alarms screamed, and the guards shouted. The X-ray machine's monitor showed a skeletal diagram of the sheriff as he passed through.

Dishevelled he emerged on the other side to see three armed security guards running towards him, their guns pointed straight at him! By now all the airport alarms were ringing. Everyone in the airport had stopped to stare at the sheriff.

"I hate kids," the sheriff groaned.

In the meantime, Mike was dashing back to the Fotoshop. If he didn't get the photos before the sheriff was off that conveyer belt, there would be no getting away from him! He reached the Fotoshop just as the last photograph fell into the processing tray. The Fotoshop man grabbed them up, put them in the envelope, and handed them to Mike just as a shadow fell across them. Mike looked up and swallowed hard.

Sheriff Hampton's angry face was pressed up against the window.

8

At Newark Airport in New Jersey, Jackie Bailey was nearly frantic with worry as she rushed down the escalators. Squeezing past passengers and between piles of suitcases, she hurried in the direction of the arrival gates.

She was nearly at the right gate when the sight in front of her made her freeze. The children were just stepping out of the plane's walkway into the arrival area. But they were *also* stepping into the glare of media coverage. There were dozens of reporters and news teams pushing and shoving, trying to get to the children.

"Mike! Steve!" she called out. "Over here!" Jackie stormed towards the gate, pushing past other passengers to get to her sons.

Tissy's mom had already steered her daughter away from the bright lights and cameras. But reporter Libby Evans and the camera crew from Jackie's TV station managed to push their way through the crowd to Pete. "Everyone wants to know how you got there," the reporter said as she shoved a microphone in Pete's face.

Pete's dad tried to shield his son, but the reporter kept repeating the question, "How did you do it?"

"On the most amazing roller-coaster ride in the world," Pete announced proudly.

"It was magic. I can show you pictures!" Mike joined in.

Woof! Woof! Mutt was there waiting for them, courtesy of a baggage handler.

"Would you let me get to my children!" Jackie shouted, elbowing the reporters out of the way. "Mike, Steve!" called Jackie as she finally reached them. "You don't have to talk to these reporters. Just keep walking and don't talk to anyone."

Then Libby cornered Jackie and shoved a microphone into her face. "What's it like raising two boys as a single mother?" she asked. The cameras caught Jackie glaring angrily at her colleague. "Do you think that has anything to do with why they ran away?" Libby continued.

Mike could tell that his mom was furious, but he didn't know whether it was with the reporters or with him.

Just then, Mike's grandfather crashed through the doors and rushed towards them. Frank Bailey quickly hurled people out of the way as he blasted through the crowd like a human cannonball.

"What are you doing here?" Jackie demanded.

"No time to talk! Let's get out of here!" Frank grabbed Mike's arm and Jackie yanked Steve along. With Mutt in tow, the Bailey gang ran out of the building to Frank's parked van. The reporters were chasing close behind.

"How did you know about this?" Jackie asked Frank.

Mike frowned. His mother didn't sound very happy to see his grandfather, even though he'd just helped get them all out in one piece.

"It's been on TV all night," Frank explained as Mike, Steve, and Jackie jumped inside the battered van. Mutt hopped in behind them, inches ahead of the reporters. Frank leapt into the driver's seat, slammed the door in a reporter's face, and started the engine.

As soon as they were clear of the airport traffic, Jackie dialled the news room on her mobile phone and started arguing with her boss.

"What's going on?" she demanded. "I show up at the airport to pick up my boys and I have to answer to Libby Evans . . . and after I asked you specifically not to cover this!"

"But we have to tell them," Mike whispered. He and Steve were in the back of the van. Mutt was lying at Mike's feet.

"Forget it!" Steve said. "Nobody'll ever listen. Who do you think's gonna believe us?"

"But it happened – "

"Just forget about it," Steve insisted and turned away. Mike watched his older brother check something in his pocket, then stare out of the van window deep in thought.

Jackie hung up on Keefe just as the van pulled in front of their apartment. They all climbed the stairs wearily, even Mutt was dragging his tail.

Mike and Steve collapsed, exhausted, on the couch. Frank sat opposite them in an easy chair.

"Gramps, we had the most amazing adventure," Mike said, beaming. "We did it. We rode the Rainbow."

Frank smiled. "I know you did."

But Jackie wasn't interested in hearing about rainbows. "What were you two thinking about?" she shouted at Mike and Steve. "I was worried sick!"

"We didn't mean to . . ." Mike began sheepishly.

Frank stood up beside Jackie. "They've got quite a story to tell. If you'd only listen – "

"Shut up, Frank," she said and paced round the room. She stopped in front of the couch where Mike and Steve sat. "Did you stop to think about anyone else? And Steven, you should know better. By the way, your school called. You're suspended for cutting class!"

Mike looked dismayed. "Mom, it wasn't his fault . . ."

"Well, you're both grounded."

Mike watched as his brother got up from the couch and walked past his mom, muttering something under his breath.

"Excuse me?" Jackie said sternly. "Did you say something to me, young man?"

Steve turned and faced her, his eyes emotionless, "I said, no wonder Dad left you."

Whack! Jackie slapped Steve hard across the face. But Steve didn't seem to feel the blow. He just stared at his mother for a long moment, then he turned and walked calmly to his room. The door slammed loudly behind him.

"Look Jackie . . . try and calm down," Frank said, barely containing his own excitement. "I know this sounds impossible, but they really have done something incredible." He reached out to take her hand.

But Jackie wasn't in the mood for listening. She snatched her hand away. "What are you talking about? This is all your fault, Frank. It's irresponsible and unfair of you to fill their heads with this magic garbage!" She sighed. "I knew you'd be involved with something like this. What did you do? Tell the boys how much fun it would be to just take off? Just like you told your son."

Frank lowered his eyes. "He said he'd come back . . ."

Jackie's expression was hard. "I wish you'd gone with him!"

Mike watched helplessly as Frank turned and walked from the apartment, the door banging behind him. His mom sank against the table.

Mike felt terrible. He wished there was something he could say to make his mother feel better. He walked towards her, not sure how she would react.

Jackie took both Mike's hands in hers. Her voice was shaky, but tender. "Michael, what you did today, do you realise how dangerous it was?"

Mike looked into his mother's eyes for a moment and then she pulled him close, hugging him tightly. "You scared me to death," she whispered.

Mike hugged her in return. "Mom, if only you could have seen the Rainbow. What a fantastic sight!"

The next morning, Mike, Pete, and Tissy were called to the principal's office. When they got there, Mr Cohen, the science teacher, met them.

"Well, let's get it over with," said Mr Cohen, leading them into Mrs Pritchard's office.

Mr Cohen stood to the side of the children listening as Mike told the principal what had happened.

Mrs Pritchard paced up and down. "I'm sure your English teacher would be impressed with your creative talents, but I'm not amused by this childish nonsense," she said.

"I guess that means you don't believe us," Mike said with a resigned shrug.

Tissy was exasperated. "But it's all true."

Mike knew they were getting nowhere. He looked at Mr Cohen. "How do you think we got there so fast?" he asked.

The science teacher raised his eyebrows, his expression unsure. "Aeroplane?" he guessed.

"If we went by plane how could we possibly have reached the farm only ninety minutes after we left school?" Tissy interrupted.

"The Kansas police reported that you called Mrs Bailey at around six-thirty in the evening," Mrs Pritchard responded, glaring at Tissy.

"Yes, but first we walked a few miles to find the

farm and then we had to get to the sheriff's office, and then – "

But the principal cut her off. "Do you want to be suspended, young lady?"

Pete nudged Mike. "Show her the pictures," he whispered.

Mike held the envelope of photos out to Mrs Pritchard. "Here, look at these."

"This is becoming ridiculous," she said, snatching the envelope and handing it straight to Mr Cohen without even glancing at it. "Go straight to your class now. You've got twenty-four hours to tell me the truth or you'll be suspended."

Tissy and Pete looked nervously at their friend. But Mike didn't have any bright ideas. At least not at the moment.

The principal shut the door to her office, shaking her head, and the three friends started glumly down the hallway. Mr Cohen lagged behind, curiously leafing through the photographs, their bright colours reflecting in his eyes.

All of a sudden, the science teacher stopped in his tracks. He thought for a moment, then went straight to his desk in the science lab and yanked open the drawer. He rummaged through it and pulled out a battered map of the United States.

After spreading the map over the top of his desk, he traced a line with his finger. The line began at Hudson Harbour, New Jersey and stretched across Pennsylvania, Ohio, Indiana, and Illinois to stop in the state of Kansas.

"Impossible," he murmured.

The teacher again examined the pictures Mike had given him. The children's forms were silhouetted against backgrounds of different colours and lights – and the colours appeared more vibrant than anything in the room around him. He didn't know whether to believe what he saw.

He tried to consider the possibilities, but soon shook his head in frustration.

"Ridiculous!" he said, staring at the pictures. Then he threw them into his desk drawer and slammed it shut.

9

Mike thought his grandfather's living room seemed much darker than usual. He sat trying to watch television news reports with Frank and Mutt. But the images on the screen kept breaking up. In fact, since they'd got back from Kansas, it felt as if everything had gone wrong.

Irritated, Mike jumped up and grabbed the television antenna. His body twisted and turned, wildly contorting into strange positions as he moved the antenna all around. But nothing he did improved the reception.

Frank began to fiddle with the knobs on the TV set. "No, that's not right. Take it back,' he said as he motioned for Mike to move further away.

Nothing could change the colour on the screen from looking dull and faded. And the way things looked on the television wasn't the only problem.

For one thing, no one believed them. And for another, Steve was acting weirder than usual. He locked the door to his room and, even though he was grounded, Mike knew his brother was sneaking out at night.

On TV, distorted pictures appeared of the four of them arriving at Newark Airport. Mike watched exasperated. "They're making fun of us!"

"Keep going, keep going back. A few more feet, go on," his grandfather continued as Mike retreated across the room.

A news report of Sheriff Wyatt Hampton flashed on screen. Static kept disrupting the sound, but Mike could still hear the sheriff lying. "My prime concern was the children's safety. As to commenting where the little monst" – said the sheriff before remembering that he was on television – "I mean the sweet, little children came from. Well, I wouldn't venture to say."

"That's it. Keep going," Frank shouted to Mike.

Mike did keep going: across the room and out of the door!

"Just a few more feet. Go on. Keep going." Frank was at last happy with the picture. "That's it, you've done it," he said, looking up from the TV set at an empty room. "Mike. Mikey. Where are you?"

"I'm in here," a voice called from the next room.

Frank went through to his bedroom to find Mike standing on his bed holding the antenna high above his head and balancing on one leg. "Gramps, when're you gonna get cable?" Mike sighed.

Frank laughed as he grabbed his grandson off the bed, and they went to join Mutt in the living room.

Mike stopped abruptly in the doorway. His own face stared back at him from the TV before the picture cut to Libby Evans in the news room.

"So that's what bored kids do for fun these days," said the reporter. "Ride Rainbows, or in adult speak, ride the airlines. Call me old-fashioned, but I remember when playing hooky meant going to the movies. Well, that's progress for you!"

Mike looked up at his grandfather with real hurt in his eyes. "I swear it's true!"

Frank placed his hands around Mike's shoulders and led him over to the sofa and sat him down. "Now let me tell you something," he said with a very serious expression. "Some people have a problem with the incredible. Like you and the Rainbow. To most folks that seems impossible. But the fact that you *did* it – "

Mike giggled. He couldn't remember seeing his grandfather try to be so serious before.

"What're you laughing at?" Frank was bemused and aware that Mike wasn't taking in a word he said.

Mike shook his head and giggled. "Go on, Gramps."

"Look, people are confused –" Frank continued.

"Like you?" Mike interrupted.

"Yeah, like – no! I mean like your ma." Frank bent down, his face close to his grandson's. "You got to look in her eyes and see what's really there because, Mikey, there's always some kinda magic going on all around."

Suddenly a wet snout nuzzled Frank's arm as Mutt's head appeared over the back of the sofa.

Frank looked from Mutt to Mike and back again. "Hey, look. The Rainbow Riders," he said and laughed.

Mike felt a little better as he left his grandfather's apartment with Mutt at his heel. So what if no one else believed them? His grandfather did. At least *he* knew they'd seen the Rainbow and ridden it! And Mrs Pritchard, well, if he couldn't convince the principal that he was telling the truth, he'd just have to make up a story she would believe –

BANG!

The sound of a gunshot echoed through the street. A moment later a police car roared by, siren shrieking, tyres squealing. Mike shivered. The air was still and heavy and things felt different than usual somehow. He didn't feel safe, and began to hurry towards home.

But Mike didn't feel any safer when he reached his apartment. It was in complete darkness and there was a strange quiet. He walked over to Steve's door, Mutt by his side.

"Steve?" Mike called out, but there was no answer. He tried the door, but it was locked. "Steve, what's going on?"

Still no reply.

Mutt sniffed at the crack underneath the door and started growling.

"Shhh," Mike said and knelt down to see what Mutt was growling at. A gold light was seeping out from under the door.

Steve heard Mike calling him, but decided to ignore his younger brother and his stupid dog. He had more important things on his mind – like the money he had got from pawning one of those pieces of rainbow gold. It was enough to buy that motorcycle – the one just like his father had ridden.

Steve fingered the two remaining gold nuggets. Their golden light radiated throughout the dingy room. The

Tigers would have to accept him now that he'd have a cool bike and enough money to throw around.

Then there'd be nothing to stop Bridget from going out with him.

The next morning, Mike was leaving for school when he heard shouting and barking from Mrs Williams's apartment. He rushed over and knocked on the door, "Mrs Williams, Mrs Williams, are you okay?"

The door flew open and there stood Mrs Williams dressed in a dull housecoat. "Mind your own business," she snarled and slammed the door in his face.

Mike stood there a moment, confused. Mrs Williams was usually so friendly. Then he shrugged and went outside.

This wasn't the best start to a day, Mike thought as he looked up at the overcast sky. It was a dirty grey. Weird formations of dark clouds moved fast across the sky. The ground was littered with rubbish.

Old Stan was leaning out of his window, but when Mike waved, Stan just slammed the window shut. He didn't even say good morning.

Mike saw a woman sitting outside Bob's Comics. She looked pale and out of breath. Mike was about to ask if he could help her when Bob stormed out of the store. "You buying or what?" he yelled at her. "Get outta here!"

Mike was shocked and confused. What was wrong with everyone? They seemed to have gone crazy!

Suddenly a noise from behind made him jump and

he spun around, but it was only the doors of the school bus as they hissed open.

He climbed aboard and was relieved to see Tissy and Pete already on the bus. He rushed over to where his friends sat.

Pete looked uncomfortable. "I don't get it. It's cold outside but it feels so hot."

"And have you seen how dull everything looks?" Mike remarked, examining his arms and hands. "Even our skin looks grey."

Tissy seemed about to answer when the school bus screeched to a stop. A red jeep raced across the junction ahead, just missing them.

"Lunatic!" the bus driver shouted.

Pete had been thrown up and nearly over the top of the seat in front of him. "Whoa!" he groaned as he righted himself.

Tissy hadn't taken any notice, she was staring out the window at something. "Hey, isn't that Steve?"

Mike and Pete looked out and saw Steve straddling a gleaming black Triumph motorcycle. He was talking to a pretty dark-haired girl in a miniskirt. Mike recognised her – she was the girl Steve had followed the other day when he had almost been beaten up.

"Where'd he get the bike?" Tissy asked.

"Where'd he get the *money* for the bike?" Pete added.

"Or the driver's licence," Tissy pointed out. "He's not sixteen yet."

"Yeah," added Pete. "Steve could get into big trouble driving that thing around."

Mike was even more surprised than his friends. He had no idea where Steve had got the bike, or the money for it, but the whole thing left Mike with a bad feeling that he'd better find out.

Mr Cohen had tried to forget about the wild story his students had told. But he just couldn't stop thinking about it.

He also couldn't stop thinking about the odd changes that had been happening ever since the kids had come back. There was a dullness to everything – even his skin – and the atmosphere was generally dismal. It seemed that people's moods had become that way too.

He frowned at the posters on the wall of his science classroom. They were brand new, yet their colours looked faded and muddy.

Again his mind returned to his three students. Automatically, he opened his desk drawer to examine the pictures they had taken.

How could they have possibly travelled to Kansas in such a short time? he asked himself for the hundredth time as he flipped through the photos. He didn't understand it. A flight from New Jersey to Kansas took four hours and nineteen minutes, so how could the children have reached the sheriff's office so soon?

As he stared at a photo of Tissy, an idea occurred to him. Before all of this began, Tissy had spent hours working in the computer room. Maybe a clue to this whole thing could be found there.

His footsteps echoed through the school hallways as

he made his way to the computer room and straight over to the terminal where Tissy had worked.

He punched at the keyboard, trying to open Tissy's files. But each time he tried, the words ACCESS DENIED flashed across the screen. Mr Cohen sighed in frustration. Tissy must have locked her files. But there had to be a password. What could it be?

He gazed around the darkened room, searching for inspiration, then he remembered the Rainbow pictures. He had carried them from his office without even realising it. Once again, he leafed through them. Their colours were impossibly strong and vibrant as ever in contrast to everything else around him.

That gave Mr Cohen an idea. He typed R-A-I-N-B-O-W. The machine released a high-pitched beep as Tissy's locked computer files sprang open. He saw a file marked TARGET, and he opened it. A map suddenly drew itself across the screen.

The teacher's eyes sparkled for the first time that day. A few minutes later he had printed out the map and was heading for his car.

He drove quickly through town until he reached the grey stone walls of industrial buildings located near the docks.

He checked the computer-generated map on the passenger seat beside him then drove over the canal bridge leading into the heart of the docks. Across the wide river, the ghostly skyline of Manhattan rose up in the distance.

He braked when he saw four children's bikes leaning up against a fence. This must be the place, he thought.

Then he noticed the huge triangular crater in the ground. He climbed over the fence and bent down to the crater. The soil looked odd. Mr Cohen scooped some up and let it gently slip through his fingers. He rubbed the dirt from his hands and stared in amazement – no, *shock* – as the colour of his skin brightened. The soil seemed to restore the pink fleshy tones to his complexion! Then it faded away!

Mike stood outside the Galaxie Diner. The place had definitely lost its sparkle. The chrome that used to be so shiny now looked dingy and tarnished. Mike stepped inside and looked around for Pete and Tissy. They had arranged to meet here after school, but Mike saw that he was the first to arrive.

The inside of the diner was even worse than the outside. The booths looked old and drab. The seat cushions were faded and every coffee stain was clearly visible. The atmosphere was no longer warm and inviting.

Only a few customers sat around, staring into their coffee cups or picking at their food. And Mike thought Pete's mom looked really tired as she sat in one of the booths teaching the businessman Spanish.

Pete's dad sat at a table at the far end of the diner, his head resting on his arms. Just then Pete and Tissy came in. Pete rushed over to his father. "Dad, are you okay?" he asked, worried.

His father looked up. "Yeah. Just tired. But I don't know why. We haven't been busy. Business is bad and people have been acting strange. Must be the weather." He gave Pete an affectionate pat on the head and went back to his work behind the counter.

Suddenly the children heard a loud rasping sound behind them. It was Pete's mother. She clung to the side of the counter, chest heaving. The businessman rushed to her side. "Someone help!" he shouted.

Pete rushed over, "Mom, you okay?"

"I need air . . . I can't breathe . . ." she clutched at her throat.

Pete's dad rushed over and put his arm round her. He half-carried her to the back of the diner.

The sound of Pete's mother's painful wheezing was shattered by the sound of glass breaking nearby. It was Bob's Comics. Vandals had thrown a bench through the store window and were climbing inside.

Mike shuddered. The neighbourhood used to be so pleasant and safe. When had things started to go wrong? he wondered. Then he froze as the answer hit him: It was after they'd ridden the Rainbow! Could all of this be connected to their Rainbow ride?

"But what did we do wrong?" Pete asked after Mike had shared his suspicions with his friends.

"Maybe we damaged the Rainbow," Tissy suggested.

A gang of bikers roared past, drowning out her voice. Mike recognised Butch from the Motorcycle Bone-house.

Suddenly the answer clicked in Mike's mind. "The bike!" He needed to find his brother. And fast.

While Mike was racing for home, Steve was about to leave. He went to the closet door and took out his father's old brown leather jacket. It had a picture of a motorcyclist riding through a line of flames and the words WALL OF FIRE inscribed in orange letters. He handled it carefully, as though it were his most prized possession.

Steve slipped his arms into the jacket and shrugged it on. It was a little big, but it fitted. He took a long hard look at himself in the mirror and smiled.

"Just like Dad," he whispered.

Then he grabbed his helmet and left.

When Mike reached the apartment he headed straight to Steve's room. This time he didn't even knock. He tried the doorknob and to his surprise it wasn't locked.

"Steve?" he called. "Are you here?"

But there was no sign of his brother.

Mutt had followed Mike into Steve's bedroom and now stood sniffing at a small wooden box on the dresser.

Mike walked over and threw back the lid.

Shining before him was a gleaming nugget of gold.

10

It was dark by the time Steve pulled into the courtyard outside the Tigers' den. He turned off the engine, propped the Triumph on its stand, and made his way into the rear of the dimly lit building. Loud music blasted from the basement as Steve walked towards the stairway leading down.

Steve noticed Bridget at the bottom of the stairs and flashed her a smile. She smiled back, then turned and disappeared into the dingy room.

It won't be long now, Steve thought as he touched the gold nugget that rested deep inside his jacket pocket.

None of the Tigers noticed Steve coming down the stairs. He stood in the doorway for a moment, letting his eyes adjust to the dark, smoky room. Ripped posters hung from a wall of peeling paint, and old furniture was scattered around haphazardly. There was a large pool table in the middle of the room. Arcade games along one wall were littered with empty beer cans and fast-food wrappers.

Steve spotted the Tiger with the leather jacket punching at a pinball machine. The other Tigers were drinking and smoking as they played pool or danced.

Suddenly, the music was turned off, throwing the basement into an eerie silence. All the Tigers turned towards Steve, their expressions changing from surprise to anger.

"Hey, what are you doin' here, Bailey?" one of the gang shouted angrily.

"Did you bring your kid brother?" another taunted.

"Hello, Bailey." The guy in the leather jacket spoke in a voice that sounded more like a threat than a greeting. He walked towards Steve, the other Tigers following his lead. "We were expecting you," he said, resting his right fist in the palm of his left hand.

Behind him the other Tigers started punching their fists into their palms, too.

Steve realised that he had made a big mistake coming here, but it was too late to get away –

The Leader spun the boom box onto full volume. The loud rock music belted out, covering up the sounds of breaking chairs and punching and kicking.

Steve left the Tiger Den cut and bruised, but at least he had fought hard enough to make it out in one piece. He leaned on the fire escape outside for support as he tried to catch his breath. Tears stung his eyes and mixed with the blood that flowed from his nose.

Steve shivered. It was chilly and one of the Tigers had taken his jacket – with the gold nugget inside.

The sound of an engine revving made him look up just in time to see one of the gang stealing his bike and riding off into the night.

Steve walked home through the deserted city. Sirens

wailed as he watched two police officers chase down a mugger.

He was only two blocks from home when he caught sight of a man and woman struggling in an alleyway. The man was trying to snatch the woman's bag but the woman clutched it tightly. Suddenly Steve thought he recognised the woman's long auburn hair and dressy clothes.

"Mom!" he screamed running towards her.

The startled mugger lashed out, knocking the woman to the ground before running off with her bag. The woman lay still.

Steve rushed over and bent down, afraid of what he would see. If that man had hurt his mother –

A terrified face stared up at him and then the woman began to scream.

"I'm sorry – I thought – " Steve said as he backed off. But the woman kept screaming. Steve heard sirens coming his way and took off, running towards home.

Jackie Bailey looked up when she heard the front door close. She saw the silhouette of her older son moving through the darkness. "Steve?" she called. "Steven, where have you been? It's dangerous out there. Besides, you're grounded, remember?"

Then she saw his bloody face.

"Omigosh, Steven, are you all right?" She rushed over to him.

"I just thought the gold would buy me everything," he said, trembling. "I thought the Tigers would want me and I'd be somebody. But it didn't work. They still didn't want me. They took my motorcycle and Dad's jacket and the piece of gold – " Steve bit his lip, thinking about the gold nugget.

"I only took a few pieces of gold, there was so much of it in the Rainbow. It was everywhere. I didn't know, I really didn't. Now look. Look what I did. It's like the end of the world," he sobbed, unable to look at his mother. "I'm sorry, I'm so sorry," Steve cried as she approached him.

Steve's mom took his face in her hands and gently stroked his bruised cheeks. "So am I," she whispered softly to him. "So am I."

Mike had been watching the whole thing from the doorway to his room. Now he came out to join his family. Mutt followed a few feet behind.

Mike knew what they had to do, and there wasn't a moment to spare. It was up to them to save the world.

Steve, Mike, Pete, Tissy, Jackie, Frank, Mr Cohen and Mutt were gathered round the dining room table in the Bailey apartment. The photos from the inside of the Rainbow were spread out in front of them along with the one remaining nugget of gold.

Frank began by pointing to the only spot of colour left in the apartment – the pictures Mike had taken. "If it's the Rainbow that gives colour to the world – "

" – and if it's the gold that makes the Rainbow work – " continued Pete.

" – then when Steve took the pieces from the Rainbow, he damaged it," Mike concluded.

"So we have to get the gold back to the Rainbow," stated Tissy.

"But how?" wondered everyone.

As the night turned into early morning, the group worked out a plan. On his laptop, Mr Cohen devised a computer program to figure out how quickly the colour was fading.

"According to my calculations," said the teacher, "we've got until four-forty-five this afternoon. That's less than ten hours from now; then all the colour's going to be gone. No colour, no photosynthesis – no photosynthesis, no oxygen," the teacher concluded.

This was even worse than Mike had imagined.

"We won't be able to breathe," said Tissy.

"Like my mom," Pete added.

Steve put a bowl of popcorn on the table and lowered himself into the seat beside Tissy and Pete. He looked down as he spoke, his voice barely audible. "I only took three pieces."

Mike exchanged a look with his friends. "How much doesn't matter," Mike said to his brother.

"Right. You broke the circuit," Tissy explained.

"We've got to get *all* the gold back and return it to the Rainbow," Mike said, leaning over the map of Hudson Harbour now spread out on the table before them.

Steve traced his fingers across the map. "The pawn

shop where I sold the first piece of gold is over here on Ninth, and the Tigers' hangout is over on Colvern."

Mike's mom grabbed Tissy's coat and thrust it at her. "C'mon Tissy. We're going to the TV station. We can use the weather computers to find the right conditions for the Rainbow somewhere."

"Gramps and I can go to the pawn shop," said Mike.

"I'll show you where the Tigers hang out," Steve said to Mr Cohen.

"I'll go with you," said Pete, grabbing his coat as he followed them out of the door.

When Jackie and Tissy arrived at the TV station, Keefe Russell was the only other person there. His shiny suit was wrinkled and stained. "Jackie. Thank God you're here. Hardly anyone showed up," he said excitedly motioning around the empty office.

Even though the news room was deserted, the computers and printers were still spewing out data. All the monitors displayed violent scenes from around the city.

Jackie tried to ignore Keefe, but when he blocked her way, she decided she'd had just about enough of this guy.

"Get out of my way, Keefe," she warned.

"No, I need you to cover the news. I need – "

"I said, get out of my way, you jerk!" exclaimed Jackie, then she lunged forward. The force of her push sent him back so hard that he slid across a desk and landed on the floor, a sheaf of paper fluttering down over him like snow.

"Hey!" he squealed, his eyes wide with disbelief.

Tissy couldn't resist the temptation, and as she walked past, she swung her bag and whacked him in the knee.

"Hey!" he squealed again, but he didn't follow them as they headed across the newsroom and straight for the weather database. Instead, Keefe quickly hauled his wounded pride off the floor and limped out of the line of fire.

When she got to the database, Jackie swiftly began to search through the sheets and sheets of printouts that listed weather conditions.

Meanwhile, Tissy sat down at a computer bank and called up the database that could connect her to the station's weather monitoring service. Soon she had one computer screen displaying a satellite picture of universal cloud cover and another showing global temperatures rising.

"Somewhere, somehow . . ." Jackie muttered as she scanned the pages, "we've got to find the right conditions for the Rainbow!"

Mike didn't think that they would ever reach the pawn shop. The van's motor was coughing and spluttering and they could barely see where they were going. Foul-smelling smoke drifted around them like a sickening fog, screaming sirens echoed among the buildings, and rubbish and broken glass littered the streets.

Most of the city had been sealed off. Wrecked and

abandoned vehicles blocked the way, their shapes hardly visible.

When they at last reached the pawn shop a sign read: CLOSED DUE TO ILLNESS, and the door was firmly locked.

"Now what?" Mike asked baffled.

"Haven't I ever told you about my skills as a great lock-picker?" Frank asked, proudly. "I did a whole Houdini routine at the carnival."

He pulled a paper clip from his pocket and straightened it out. "All I need is this piece of wire. Don't worry, we'll be inside in a few seconds."

Mike looked around nervously as Frank fumbled with the lock. Soon sweat began to break out on Frank's forehead. But at last there was a loud click and Frank turned the knob with relief.

BBBBBBRRRRRRRRRRRING!

"Oh no, you set off the burglar alarm!" exclaimed Mike. "Now what do we do?"

"We move fast!" Frank shouted above the noise. He burst through the door, Mike and Mutt right behind him.

The pawn shop was cluttered with old musical instruments, furniture, and a ton of other weird stuff that Mike had never seen before. Junk was piled up almost to the ceiling.

"This is like trying to find a needle in a haystack!" Frank shouted over the clatter of the alarm.

Then Mike noticed Mutt nuzzling at an old rusty tin

and rushed across. "Over here, Gramps," he shouted at Frank.

His grandfather darted round the table excitedly. "Has he got it?"

"I'm not sure," said Mike.

The dog stood on the table, nose pressed hard against the tin, trying to force it open.

"Is that it?" Frank insisted.

"I think so," Mike replied as he snapped the lid back. But the gold wasn't inside. Just a tray full of old jewellery.

Mike's face fell. "It's not here."

But then he noticed a light shining from beneath the tray and Mike wrenched it back. There, among scraps of metal and broken watches, was the glowing golden nugget. It lit up the dark room, its golden light reflecting off their clothes and faces. They both laughed with relief.

"You did it!" Frank yelled. "C'mon, let's get outta here."

They tore out of the pawn shop and leapt into the waiting van.

Mike smiled. It had been easier than he thought. No problem, just like his grandfather said –

The sudden sound of sirens made him take back his thoughts. "Come on, Gramps, let's go!" Mike urged. If the police found them there, they'd be arrested. Then they'd never get the gold back into the Rainbow in time.

Frank kept turning the key in the ignition. The van sputtered, but the engine refused to turn over.

"C'mon, not now," Frank pleaded with the old blue van as he stroked the dashboard. "Please, not now!"

"I can see the police car!" Mike cried. "Hurry!"

Frank turned the ignition key again, frantically trying to start the engine. Finally, the ignition fired, the engine burst into life, and Frank thrust the van into gear. He hit the accelerator just as the police car turned the corner.

The van sped away and out into the street.

11

Mr Cohen pulled his car to a stop in the courtyard outside the Tigers' hangout.

"What's the plan?" Pete asked.

"What d'you mean? They're just kids, right?" said Mr Cohen with a shrug.

Steve gave the teacher an uneasy look.

Loud rock music blared out from the basement.

"Just a bunch of – " The teacher stopped and his eyes grew wide when he looked through the window.

In the room below, the Tigers stood round a table watching as one of their members played a dangerous game. The mean-faced kid was stabbing a knife between each of his fingers. The knife moved faster and faster as the music reached a crescendo. The gang members watched intensely until the mean-faced Tiger suddenly slammed the knife down into the wood. A gasp of relief could be heard around the table.

" – uh, yeah . . . just kids . . ." Mr Cohen whispered.

Pete looked at him. "So what was that plan again?" he asked.

They could see the gold nugget – it was on top of the pool table in the middle of the room. But there was no way they'd be able to get through the Tigers to reach it.

The teacher opened his mouth, but no sounds came

out. Then he noticed the ventilation shaft nearby. It led straight down into the basement room. Mr Cohen tested it with his fingers. "You know what these guys need?" he asked with a sly little smile. "A science lesson."

The teacher went back to his car and unlocked the trunk. Inside sat a bunch of different items, among them was a small tank filled with helium gas. Mr Cohen took a plastic pipe, connected it to the cylinder, fed it through the ventilation grill, and turned the helium tank's valve to open.

Then they waited.

After a short while, one of the Tigers got up and moved over to a pretty blond. "Hey, you wanna dance with me?" he said in a squeaky chipmunk voice. The kid sounded just like Pete had when he'd been goofing around with the helium in science class.

Another Tiger jumped up. "You talk stupid, man!" But he sounded exactly the same. Squeaky.

The other Tigers laughed and pointed at him.

"You dissing me?" the Tiger squeaked.

"What you gonna do about it?" the other one replied in a high girlish voice.

Soon they were all pushing and shoving each other!

Mr Cohen and Pete laughed as they watched from the window until they noticed that Steve was missing.

"Steve?" The teacher looked about frantically. "Where's Steve?" When they looked back through the window, the pushing and shoving had escalated and the whole gang was fighting. And Steve was sneaking down the stairs towards the rioting Tigers!

"Oh, no, Steve!" Pete exclaimed.

They watched as Steve darted around unnoticed. He dived under a pinball machine as bottles and chairs smashed around him. Next he carefully crawled along the floor towards the pool table. The Tigers continued to punch and kick each other, still oblivious of him.

Crouching under the pool table, Steve reached his hand over the top and began searching blindly for the piece of gold. He felt across the drab greenish surface and past the pale pool balls. His fingers were inches away from the nugget, when one of the Tigers saw him and snatched it away.

Steve saw the shadow fall across him and stood up. He was face to face with the mean-looking Leader. The kid was even wearing the Wall of Fire jacket that had belonged to Steve's dad.

"The gold's mine now," said the Tiger with steely determination. "Like your jacket and your bike."

"It doesn't belong to you!" shouted Steve, lunging forward.

But the Tiger wasn't sticking around. In a flash he fled up the stairs throwing stacked chairs behind him to block Steve's path.

Steve chased after him, tossing the obstacles out of his way as he took the stairs three at a time. He raced after the boy, following through the decayed building and out into pale grey daylight.

By the time Steve reached the archway, the Tigers'

Leader was already halfway down the block and heading for a motorcycle. It was Steve's stolen Triumph.

Steve took a deep breath, remembering everything his track coach had ever taught him. Then he bolted after the Tiger like an Olympic sprinter fresh off the starting block. He doubled his efforts, straining with all his might as he tried to close the gap.

The Leader was fast, but Steve was faster. He was ten yards away and closing when the Leader darted round an abandoned pickup truck.

Steve saw his chance and, timing his move perfectly, jumped onto the truck and leapt off it like a jumper off a springboard. He flew through the air and landed right on his target, bringing the boy down with a thump.

The Leader groaned as Steve flattened him and the golden nugget rolled from the Tiger's palm. Steve quickly grabbed it up. Then he tore his Wall of Fire jacket off the still-stunned Leader's back.

Steve slipped the jacket smoothly over his shoulders, straddled the Triumph, and kicked the motor into action.

He rode past Pete and Mr Cohen, tossing them a smiling thumbs-up. The two grinned in return then hurriedly piled into Mr Cohen's car and followed Steve home.

Mike sat on the sofa next to his brother. The golden glow from the three nuggets shone on their faces and lit up the room.

Frank paced up and down while Mr Cohen and Pete

waited nervously at the dining room table. No one spoke as they waited for Jackie and Tissy to return.

Suddenly the front door flew open and Jackie and Tissy ran into the living room. Jackie held out a piece of paper. "Bad news. We looked for signs of rainfall for a rainbow, but the only place with the right weather conditions is Buffalo." She threw the paper onto the table. "We'd never get there in time."

They all looked at each other silently for a moment.

"It may already be too late," Mr Cohen said as he closed his laptop.

Mike took in their defeated expressions, thinking hard. There must be something they could do. And then he saw Mutt sitting in the doorway and it hit him. "No, it's not too late," he said as he shot to his feet. "I can find the Rainbow!" Mike grabbed the three pieces of gold and put them in his backpack.

"What do you mean, Mikey?" his mom asked.

"Don't you see? We've been too scientific . . . we've lost the magic," he replied.

Mr Cohen shook his head, not understanding. "You're not making any sense."

"Yes he is," said Jackie, walking over to her son.

"The Rainbow came to me before," Mike said, looking at their mystified expressions. "I think it will come again."

He looked at his watch as he crossed the living room towards his brother. "C'mon, Steve, it's four-fifteen. That means, according to Mr Cohen's calculations, we've only got thirty minutes before all the colours are gone. I'll need a ride to get to the railroad yard in time."

"No problem," Steve replied and followed his younger brother from the apartment.

Steve drove his Triumph as fast as he could through the drab hostile streets. Mike held on behind him. Mutt ran alongside, finding shortcuts over low walls and through holes in fences.

Turning a corner at the bottom of a hill, the boys heard shouting. They noticed huge clouds of smoke coming from the park that separated them from the railroad yard.

"Uh-oh," Steve murmured as he brought the bike to a quick stop. Mike took in the awful sight. The park was a sea of rioting people. On one side, the rioters had pinned police officers up against a giant statue. And on the other, smoke billowed from the burning hulks of cars creating a white wall of fire.

"What do we do now?" Mike asked. This was the only direct route to the railroad yard. If they drove round the park, they'd never make it in time.

Steve just stared ahead, fixated by the wall of fire. "Brace yourself, Mikey!" Steve yelled and gunned the engine.

Suddenly a man appeared from nowhere, swinging a baseball bat at them. Mutt leapt through the air and caught the man's sleeve in his jaws, bringing him down in a flash of teeth.

Mike and Steve shot by the falling man and over a

ramp. Mike watched the wall of fire grow nearer and nearer, but instead of fear, he felt a strange comfort as he clung to his big brother's back. Maybe it was just the old leather jacket Steve was wearing, maybe something more, but Mike felt as if his dad were sitting in front of them in that moment, protecting them from the burning heat.

A second later they skimmed through flames and out to the far side of the park.

"You did it, Steve!" called Mike. "You jumped the Wall of Fire! Just like our dad."

"Yeah, I did, didn't I?" said Steve, a smile touching his face. "Like Dad. I did it."

A few minutes later, Steve brought the bike to a stop in the railroad yard. Mike scanned the area around him, looking for any trace of the Rainbow. There was only a smear of colour left on the graffiti-covered boxcars and in a minute even that dissolved before their eyes.

They got off the motorcycle and gazed round at the bleakness. There was no sound, no movement, no sign of the Rainbow.

"Where is it?" Steve asked.

"Here," Mike answered. "It has to be!" He glanced at his watch. "We're almost out of time."

Mike ran across the old railroad tracks to where Mutt sat on the edge of the shallow crater. Staring up at the dark barren sky, Mike whispered, "I don't know where it is, Mutt."

The second hand of Mike's watch ticked around.

Thirty seconds, twenty-nine, twenty-eight, twenty-seven –

Mutt began to bark.

"Please . . . I know you're up there, Rainbow," called Mike. "I can feel it . . ."

Steve looked on uneasily as Mutt kept barking and Mike searched the darkening skies.

Mike closed his eyes tight, then opened them hopefully. Still nothing. The seconds passed and there was total silence in a colourless world.

Then Mutt cocked his head and looked skywards. A ripple of wind brushed the dog's fur as a low rumbling noise sounded in the distance. It rolled steadily towards them, building to a loud roar. The wind howled.

Suddenly there was a deafening blast! The sky cracked open and an ugly brown light thundered down into the crater!

It was the Rainbow. But not the beautiful multi-coloured creation that Mike had seen before. It was a putrid red-brown. The Rainbow looked sick, like it was dying.

Mutt raced towards it.

"Mutt! Don't – " Mike called, but it was too late. The dog leapt into the Rainbow and disappeared.

Mike was only inches from the Rainbow, standing under its dark shadow. He looked down at his hands which disintegrated into tiny particles, as if being absorbed by the Rainbow.

Then he heard his brother scream, "Mikey, no!"

But the Rainbow had already engulfed him in a

crackle of electricity. For the second time, Mike was inside the Rainbow.

12

Mike's stomach churned as he felt himself pulled by invisible hands. He zoomed up through the brown Rainbow and saw that Mutt was already flying up ahead of him.

The Rainbow tugged Mike higher and higher. He felt a cold wind rushing past him, battering his face and eyes, pulling his muscles taut. He had to squint through the mud-brown mist to make out the buildings of Manhattan and the landscape that rapidly disappeared far below.

Mutt flew higher, too, zooming far into the distance as Mike struggled to grab his backpack and free the gold.

The forces were incredible. Mike could hardly move his hands as he travelled through the Rainbow, spinning through the menacing, howling vortex.

This sad, sickened Rainbow was frightening to Mike, but he tried to keep calm. He had to get the gold out of his pack!

The forces pulling and turning his body were almost unbearable. Somehow he managed to bring the bag closer to him, but the pressure was too much and it shot away.

He strained hard to bring it back again, and eventually clutched it close to his chest. Mike tugged at the zipper, trying to pull it across, and reached inside for the gold.

With a mighty thrust, Mike threw the gold pieces into the Rainbow. To his amazement they flew far ahead of him, fitting themselves into part of the centre of the Rainbow.

Mike remembered the Rainbow's centre as being brilliant, warm, and dazzling. Now it was dark and cold, black and terrible.

The three golden nuggets that flew from Mike's hand now formed a triangle. A bright spark of light jumped from it, lighting the next golden nugget and the next and the next.

The connection had been made, thought Mike. Now that the missing links were replaced, the Rainbow was beginning to heal!

Flashes of gold spread before him, creating a fantastic spider-web of light. The magical effect triggered connections inside the whole Rainbow. Gold beams flashed in every direction, lighting up the bubbles of energy with brilliant colours.

Mike's worried, strained expression was transformed into one of amazement as he watched the glowing colours sweep the entire length of the Rainbow. With a blinding flash and a loud sonic boom, the Rainbow began to re-energize!

After Steve and Mike had left, Jackie took the others up to the roof of the TV station. It was one of the highest buildings in Hudson Harbour and had a good view of the entire city.

At first all they could see was smoke rising from the ruins around them. Then, suddenly, they saw the sick, swirling Rainbow flash brown across the sky. The group stood among the satellite dishes and antennae nervously waiting and wondering what would happen next.

The only visible colour was the blue of Jackie's eyes as she stared at the brown Rainbow.

Then a few raindrops began to fall. To everyone's astonishment, every drop that touched them brought colour to their drab grey skin. As drops splashed onto their clothes and their hands and faces, the colours began to return.

"You did it, Mikey. You did it!" Frank yelled, jumping up and down as the colouring rain cascaded over them.

Soon every building, flower, and tree around them had returned to its own special shade and hue. And every person's own unique colour had returned to his or her skin. Every colour, every shade was glorious to see again!

Jackie looked out at the city that had been grey and dead and was now rich and vibrant with colour and life. It thrived under the shelter of the multicoloured glow of the majestic Rainbow, which arched far into the clouds and out of sight.

Frank stretched out his hand, offering it to her. She took hold and squeezed it tightly, smiling with joy as they both turned to face the Rainbow where Mike spiralled thousands of feet away.

Meanwhile, back at the railroad yard, Steve was looking at his hand as the first few drops of rain fell. He watched, amazed, as the pink fleshy tones returned to his skin.

Soon the isolated drops grew into a steady shower and then a curtain of rain. Steve arched his head back and let the water cascade over his face, restoring the natural shades.

With pride and joy, he punched a fist into the air as the colour flooded back to the ground and the sky, even back to the graffiti-covered boxcars nearby.

"Way to go, little bro!" shouted Steve to the sky.

Inside the Rainbow, Mike watched as the different colours rose and swirled before him. Red appeared, then came orange, then yellow. Now green. Finally blue and violet sprang up before his eyes as the full spectrum of the Rainbow's colours was released shade by shade.

The glorious colours then swept through the vast arc in a dazzling kaleidoscope. Beams of glowing gold raced through the Rainbow, banishing the last bits of darkness.

The glowing centre of the Rainbow was just ahead. It seemed to signal to Mike like a beacon, and he floated towards it, surfing along the glowing light.

The centre of the Rainbow was its most powerful part, where primary colours combined to form the purest white light. Just a few moments ago it had been black and terrible, sickened by the missing nuggets of gold. Now it was whole and bright again, and Mike's eyes were wide with wonder as dazzling particles seemed to bounce off him and dance all around.

The Rainbow actually seemed happy. It was playing with him, realised Mike. It was thanking him!

After a few minutes of tumbling and rolling and laughing with joy, Mike felt as though the Rainbow was urging him forward again. In front of him, Mike soon saw a stream of colours that flowed like water to form a multicolored bridge to earth.

That's when he noticed the dip ahead. The point where the Rainbow started to head downwards. He gulped as he began the descent. He felt as if he were free falling as he plunged down through the clouds, Mutt not far ahead.

When he emerged on the other side of the Rainbow, he gasped at the sight below. A vast brightly coloured landscape stretched as far as he could see.

Purple mountain peaks rose above deep green valleys. A curling blue river fed the red, orange, and yellow blossoms growing by its banks. The warm, brilliant colours drenched his eyes and made him grin from ear to ear. The land seemed to glow, as if it were happy too – the colour had returned to the world!

As he continued to descend inside the Rainbow, Mike noticed that the trees below him weren't part of just any woods. This place was a huge forest, and its trees and plants were the size of houses!

Mike became dizzy with spinning and rolling and

looked at the ground as it grew closer. The shapes quickly became more distinct. He floated through low-lying mists and down past a canopy of leaves as the Rainbow gently lowered him to the ground.

As Mike sat stunned on the soft earth next to Mutt, a low rumbling noise began. In a flash, the Rainbow whipped upwards, its multicoloured bands swallowed by the clouds and disappearing from view.

Mike stared at the sky for a long moment, then he looked at his surroundings.

Thick, spiky plants grew next to shrubs with brilliant coloured exotic flowers. The leaves were of every possible shape and size, long-pointed ones, short-rounded ones, umbrella shaped ones, growing in all shades of green.

"Mutt, I don't think we're in Kansas this time!" he exclaimed.

Actually, it looked exactly like the picture of the Amazonian rainforest in his geography textbook.

He heard strange sounds – buzzes, hisses, howls. The shrill cry of a bird gave way to the squeal of monkeys. Mike heard loud rushing water and when he turned he saw the most amazing waterfall.

Mike's eyes widened with wonder. There were plants and flowers he'd never seen before and strange looking creatures and beautiful animals in nearly every tree.

Two multi-coloured parrots were perched on a tree nearby, and a long-haired monkey peered out of the vegetation at him. It looked just as startled as Mike!

Mike laughed. His eyes were bright and his face glowed with excitement as he turned to his dog.

"Mutt?"

The dog looked back at him.

"Well, boy, guess it's another long walk home."

Mike's face broke into a huge smile. He realised that he wasn't scared anymore. He had Mutt, and after all, it was another adventure!

Other great reads from ***Red Fox***

Further Red Fox titles that you might enjoy reading are listed on the following pages. They are available in bookshops or they can be ordered directly from us.

If you would like to order books, please send this form and the money due to:

ARROW BOOKS, BOOKSERVICE BY POST, PO BOX 29, DOUGLAS, ISLE OF MAN, BRITISH ISLES. Please enclose a cheque or postal order made out to Arrow Books Ltd for the amount due, plus 75p per book for postage and packing to a maximum of £7.50, both for orders within the UK. For customers outside the UK, please allow £1.00 per book.

NAME____

ADDRESS_____________________________________

____

Please print clearly.

Whilst every effort is made to keep prices low, it is sometimes necessary to increase cover prices at short notice. If you are ordering books by post, to save delay it is advisable to phone to confirm the correct price. The number to ring is THE SALES DEPARTMENT 0171 (if outside London) 973 9000.

Humour and Adventure in the Redwall Series

Read the REDWALL adventures by master storyteller Brian Jacques – an award-winning bestselling series of epic adventures based around Redwall Abbey, home to a community of peace-loving mice.

REDWALL

As the mice of Redwall Abbey prepare for a feast, unknown to them, Cluny, the evil one-eyed rat, is preparing for almighty battle . . .

ISBN 0 09 951200 9 £4.99

MOSSFLOWER

The gripping tale of how Redwall Abbey was established through the bravery of the legendary mouse, Martin.

ISBN 0 09 955400 3 £4.99

MATTIMEO

Slagar the Fox is intent on bringing destruction to Redwall Abbey and the fearless mouse warrior, Matthias. He'll stop at nothing – including Mattimeo, Matthias's son

ISBN 0 09 967540 0 £4.99

MARIEL OF REDWALL

A young mousemaid is washed ashore on the fringes of Mossflower country. Battered and bruised, she makes her way to Redwall Abbey, where the story of her horrific ordeal unfolds.

ISBN 0 09 992960 0 £4.99

SALAMANDASTRON

Why did the sword fall from the Abbey roof? Who is the white badger? And can the good creatures triumph over Ferahgo the Assassin?

ISBN 0 09 914461 5 £4.99

MARTIN THE WARRIOR

Bedrang the Tyrant stoat holds many creatures prisoner in his fortress on the coast. But a young mouse, Martin, refuses to obey the Evil Lord, and plans a daring escape.

ISBN 0 09 928171 6 £4.99

Humour and Adventure in the Redwall Series

THE BELLMAKER

Bar your doors! The Foxwolf and his horde of grey rats are coming to Southsward! After dreams of Martin the Warrior, old Joseph the bellmaker sets off with four young warriors to help friends abroad . . .

ISBN 0 09 943331 1 £4.99

THE OUTCAST OF REDWALL

Ferret Swart Sixclaw and arch enemy, Sunflash the Mace, swear to carry out a pledge of death – upon each other. Meanwhile a young creature is cruely banished from the Abbey – and a mousemaid finds her destiny linked to . . . the Outcast of Redwall.

ISBN 0 09 960091 9 £4.99

and coming soon in Red Fox . . .

THE PEARLS OF LUTRA

Far away, on the Isle of Sampetra, Emperor Ublaz, known as Mad Eyes, sends his fearsome lizards on a murderous mission to Redwall. Meanwhile, the abbey dwellers frantically race against time to unravel a fiendishly difficult set of riddles . . .

ISBN 0 09 963871 1 £4.99

Other great reads from ***Red Fox***

Top teenage fiction from Red Fox

PLAY NIMROD FOR HIM Jean Ure

Christopher and Nick are each other's only friend. Isolated from the rest of the crowd, they live in their own world of writing and music. Enter lively, popular Sal who tempts Christopher away from Nick . . .

ISBN 0 09 985300 0 £2.99

HAMLET, BANANAS AND ALL THAT JAZZ
Alan Durant

Bert, Jim and their mates vow to live dangerously – just as Nietzsche said. So starts a post-GCSEs summer of girls, parties, jazz, drink, fags . . . and tragedy.

ISBN 0 09 997540 8 £3.50

ENOUGH IS TOO MUCH ALREADY
Jan Mark

Maurice, Nina and Nazzer are all re-sitting their O levels but prefer to spend their time musing over hilarious previous encounters with strangers, hamsters, wild parties and Japanese radishes . . .

ISBN 0 09 985310 8 £2.99

BAD PENNY Allan Frewin Jones

Christmas doesn't look good for Penny this year. She's veggy, feels overweight, *and* The Lizard, her horrible father has just turned up. Worse still, Roy appears – Penny's ex whom she took a year to get over.

ISBN 0 09 985280 2 £2.99

CUTTING LOOSE Carole Lloyd

Charlie's horoscope says to get back into the swing of things, but it's not easy: her Dad and Gran aren't speaking, she's just found out the truth about her mum, and is having severe confused spells about her lovelife. It's time to cut loose from all binding ties, and decide what she wants and who she really is.

ISBN 0 09 91381 X £3.50

Other great reads from ***Red Fox***

Teenage thrillers from Red Fox

GOING TO EGYPT Helen Dunmore

When Dad announces they're going on holiday to Weston, Colette is disappointed – she'd much rather be going to Egypt. But when she meets the boys who ride their horses in the sea at dawn, she realizes that it isn't where you go that counts, it's who you meet while you're there . . .
ISBN 0 09 910901 8 £3.50

BLOOD Alan Durant

Life turns frighteningly upside down when Robert hears his parents have been shot dead in the family home. The police, the psychiatrists, the questions . . . Robert decides to carry out his own investigations, and pushes his sanity to the brink.
ISBN 0 09 992330 0 £3.50

DEL-DEL Victor Kelleher

Des, Hannah and their children are a close-knit family – or so it seems. But suddenly, a year after the death of their daughter Laura, Sam the youngest son starts to act very strangely – having been possessed by a terrifyingly evil presence called Del-Del.
ISBN 0 09 918271 8 £3.50

THE GRANITE BEAST Ann Coburn

After her father's death, Ruth is uprooted from town-life to a close-knit Cornish village and feels lost and alone. But the strange and terrifying dreams she has every night are surely from something more than just unhappiness? Only Ben, another outsider, seems to understand the omen of major disaster . . .
ISBN 0 09 985970 X £2.99

Other great reads from ***Red Fox***

Share the magic of The Magician's House by William Corlett

There is magic in the air from the first moment the three Constant children, William, Mary and Alice arrive at their uncle's house in the Golden Valley. But it's when they meet the Magician, William Tyler, and hear of the Great Task he has for them that the adventures really begin.

THE STEPS UP THE CHIMNEY

Evil threatens Golden House in its hour of need – and the Magician's animals come to the children's aid – but travelling with a fox brings its own dangers.

ISBN 0 09 985370 1 £2.99

THE DOOR IN THE TREE

William, Mary and Alice find a cruel and vicious sport threatening the peace of Golden Valley on their return to this magical place.

ISBN 0 09 997390 1 £2.99

THE TUNNEL BEHIND THE WATERFALL

Evil creatures mass against the children as they attempt to master time travel.

ISBN 0 09 997910 1 £2.99

THE BRIDGE IN THE CLOUDS

With the Magician seriously ill, it's up to the three children to complete the Great Task alone.

ISBN 0 09 918301 9 £2.99